Contents

Peter Townsend

Editors
Will Adams
Julia Thorley

Production
Michael Sanders
Louise Townsend

Photographic and written contributions for possible publication in *Past and Present* are welcome. All submissions should be clearly marked with the contributor's name and address and accompanied by a stamped, self-addressed envelope; written material must be typewritten double spaced. All material must be submitted on the understanding that such submission is entirely at the contributor's risk and Past & Present Publishing Ltd cannot be held responsible for loss or damage, however caused.

Contributors should also ensure that all copyrights are cleared in advance of submission, and that copyright clearance and other permissions required from third parties have been received in writing.

Material should be sent to the Editors at 2 Denford Ash Cottages, Denford, Kettering, Northants NN14 4EW (Will Adams), or 191 Kingsley Avenue, Kettering, Northants NN16 9ET (Julia Thorley).

Past and Present is published quarterly. Vol 1 No 3 will be on sale on 10 December 1993.

Annual subscriptions: UK £24.00, Europe £26.00, elsewhere £26.00 surface, £31.00 air mail, including postage and packing. Please send cheques/Postal Orders made payable to Past & Present Publishing Ltd to Unit 5, Home Farm Close, Church Street, Wadenhoe, Peterborough PE8 5TE.

Registered Office: The Trundle, Ringstead Rd, Great Addington, Kettering, Northants NN14 4BW.

ISSN 0967-8751

Pastword
Julia Thorley

Do you look at the lot of today's youngsters and feel more than a little envious of the opportunites they have? Or do you look back on your childhood and feel glad that you were born when you were?

Years ago, you may cry, we were able to run about in the countryside without fear. We made our own fun with hopscotch, football in the streets, unbothered by traffic - we didn't need TV and constant music to keep us amused. On the other hand, the coming of the Welfare State has meant that, in theory at least, there are now more and fairer opportunities for education; no one has to leave school at 14 to support the rest of the family, and the standard of living is generally higher today. Chances are you probably feel a little sympathy with both of these views.

One thing the team here at *Past and Present* is determined about is that this journal should not become a rose-tinted look back at 'the good old days', and this issue's articles provide plenty of material to support the idea that while much from the past is good, it is not necessarily better just by virtue of its age. Nor is the present always an improvement; it is just *different*, and that is what makes nostalgia so fascinating, discovering what has changed and why, whether it to be to the good or a backward step. Whatever feelings are evoked by recalling memories of your childhood, there is sure to be something in *Past and Present* that will strike a chord.

For all the Game Boys and other pieces of electronic wizardry that currently bleep and whir throughout the playgrounds of the land, it is, perhaps, reassuring to learn that one enduring piece of entertainment continues to delight children everywhere - the Punch and Judy show. So join the crowds on page 98 and discover that while many 'professors' still perform the traditional story, with all its sinister undertones, there is a new breed who have brought the story up to date to make it more relevant to today's audiences without it losing its original charm.

Boys of all ages - and probably a few girls! - will be scurrying to their lofts to discover if they have any priceless relics of their childhood languishing in a forgotten corner after reading all about collecting toy cars on page 95. If you are currently buying toys for a child or grandchild, it might be worth your while buying an extra one for yourself to keep carefully in its box as a future investment. You never know. . .

The lovely school photo comparison on page 86 will I am sure raise a few wry smiles. Maybe there is a certain amount wrong with today's education system, but one ex-pupil finds a lot to be pleased about when he revisits his old school. And if you went to on to higher education, our peep into the corridors of Bristol University will fascinate, I am sure.

Sadly, one aspect of childhood which has not changed for the better is that there is still a very real need for the good works of many charities. Barnardo's continues the vital work begun by the good doctor in 1866. Some of the problems it tackles, such as homelessness, are the same ones that faced him all those years ago, but I spent a fascinating afternoon with Director Roger Singleton, who explained to me how Barnardo's is building on its past to help children and young people - and their families - face up to the challenges and overcome the difficulties of the 1990s. They no longer run the orphanages with which their name is still, for many people, synonymous, but they continue to offer help and support in a variety of ways.

Our prize competition offers the opportunity for everyone to dig into their memories and recall the children's classics they had read to them, or which they have read to others. Surely high on the list of literary memories will be the stories of Thomas the Tank Engine and his friends, currently enjoying yet another wave of popularity with today's youngsters, thanks to the continuing publication of the stories and their transfer to the small screen via numerous video collections. Ten lucky winners will receive sets of 'Thomas' stories signed not only by their creator, the Reverend W. Awdry, but also by his son Christopher, for whom the stories were originally created and who has taken up the mantle of author in recent years. Enjoy the story of how Thomas came to be, then turn to page 121 for your chance to enter this exciting competition.

Whether you were a child or an adult during the 1950s, the idea of a trip to the dentist no doubt filled you with, if not terror, at least trepidation. On page 114 you can sit back in your chair and read how life in the surgery has definitely improved over the years. ●

If this issue stirs any particular memories - and I am sure that it will! - do drop us a line and share them with us.
And don't forget to look out those old photographs, take their 'present' equivalents, and share your reminiscences with us.

We look forward to hearing from you.

Tankerton revisited

Tankerton is a quiet resort lying to the east of Whitstable in Kent, the oyster port known to the Romans. Sadly, the oyster fishing was affected by the storms of 1953, which caused severe pollution. It was while I was on holiday at Westbrook near Margate with friends that we visited Tankerton, but that was in 1950 and I was only six years old at the time, so my memories of the trip are rather faint. However, I recently came upon some old photographs that show Tankerton Circus as it was then. In those days the town was flourishing, because it was only possible to take holidays in the UK owing to the devastation caused throughout Europe by the Second World War.

Revisiting the same spot in December 1992 I found the view much the same, although the present-day roundabout is much larger and the bold 'KEEP LEFT' instruction has been replaced by arrows and chevrons. Essentially, however, the scene is largely unchanged, except that some of the shops and shop fronts have changed use. Interestingly the site of 'The Tankerton Circus Pharmacy' of 1950 is today still occupied by a chemist. ●

Both photographs and notes by Allan Mott

Remembering...
Leeds trams
Alan Bennett

'There was a point during the Second World War when my father took up the double bass. To recall the trams of my boyhood is to be reminded particularly of that time.'

It is around 1942 and we are living in the house my parents bought when they got married, 12 Halliday Place in Upper Armley. The Hallidays are handily situated for two tram routes, and if we are going into town, rather than to Grandma's in Wortley, the quickest way is to take a number 14. This means a walk across Ridge Road, down past the back of Christ Church (and Miss Marsden's the confectioner's) to

Stanningley Road. Stanningley Road is already a dual carriageway because the tram tracks, running down the middle of the road, are pebbled and enclosed by railings, so splitting what little traffic there is into lanes. The Stanningley trams are generally somewhat superior to those on other routes, more uphol-stered, and when the more modern streamlined variety comes in after the war, you are more likely to see them on this route than elsewhere. But the draw-

A well-known 'picture postcard' scene of Leeds city centre. This is the top of Lower Briggate on Midsummer Day in 1955 with 'Horsfield' car no 235 about to turn into Duncan Street on its way to Meanwood. *A. K. Terry*

Looking north towards Leeds's main shopping area, this is Lower Briggate today. Some redevelopment has taken place over the years and, as can be seen on the left, more is currently in progress. It is pleasing that the old buildings now surrounded by scaffolding are to retain their solid facade - this would certainly never have happened in the 'pull everything down and put up a glass box in its place' '60s and '70s - some things have changed for the better! Good also to see that the splendid building on the right has been cleaned and remains to dominate the sky-line. *F. Whiley*

back with the Stanningley Road trams is that they are coming down from Bramley or even Bodley, and are always pretty crowded, so more often than not we go for the other route, the number 16, which means walking up Moorfield Road to Charleycake Park and Whingate Junction.

This being the terminus, the tram is empty and as likely as not waiting, or, if we've just missed one, the next one will be already in sight, swaying up Whingate. We wait as the driver swings outside and with a great twang hauls the bogey over ready for the journey back, while upstairs the conductor strolls down the aisle, reversing the seats before winding back the indicator on the front. The driver and conductor then get off and have their break sat on the form by the tramstop, the driver generally older and more solid than the conductor, or, I suppose, the conduc-tress, though I don't recall conductresses coming in until after the war.

Dad is a smoker so we troop upstairs rather than going 'inside', the word a reminder of the time when upstairs was also outside. On some trams in 1942 it

still is, because in those early years of the war you can still find the odd open-ended tram. We wedge ourselves in the front corner, an unexpected treat to be exposed to the wind and weather, and also an antidote to the travel sickness from which both my brother and I suf-fer, though I realise now that this must have been as much due to all the smok-ing that went on as the motion of the tram itself. Neither of us actually is sick, but it's not uncommon and somewhere on the tram is a bin of sand just in case.

So the four of us, Mam, Dad, my brother and me, are ensconced on the tram sailing down Tong Road into town or, if we are going to see Grandma, who lives in the Gilpins, we get off half way at Fourteenth Avenue.

Around 1942, though, we come into the double bass period when some of our tram journeys became fraught with embarrassment. Dad is a good amateur violinist, largely self-taught, so taking up the double bass isn't such a big step. He practises in the front room, which is never used for anything else, and I sup-pose because the bass never has the tune, it sounds terrible; he sounds as if

Left and above Alan Bennett remembers Whingate terminus, close by the quaintly named Charleycake Park, a small oasis of greenery in a heavily built-up area. Here we see Railcar 602, a truly magnificent vehicle that looked regal in its special livery of royal purple, cream and gold leaf lining. In its early life in 1953 and 1954 it visited many parts of the city on various routes, but finally settled down to work the No 25 from Swinegate to Hunslet. It is seen here in 17 June 1954 at Whingate terminus. It survived until the end of the system and was subsequently preserved and went to National Tramway Museum in Crich, Derbyshire (see page 125), where it is in working order. *A. K. Terry*

Time has truly passed by what was Whingate tram terminus. All the property remains, although now in brighter colours than the ubiquitous browns and greens of the 1950s - even the tram wire support pole still survives, now giving service as a lamp post. *F. Whiley*

he's sawing, which he also does, actually, as one of his other hobbies is fretwork. Though the instrument is large, the repertoire is small except in one area: swing. Until now Dad has never had much time for swing or popular music generally. His idea of a good time is to turn on the Home Service and play along with the hymns on Sunday Half Hour, or (more tentatively) with the light classics that are the staple of Albert Sandler and his Palm Court Orchestra. But now with Dad in the grip of this new craze, Mam, my brother and I are made to gather round the wireless, tuned these days to the Light Programme, so that we can listen to dance band music.

'Listen, Mam. Do you hear the beat? That's the bass. That'll be me.'

Dad has joined a part-time dance band. Even at eight years old I know that this is not a very good idea and just another of his crazes (the fretwork, the home-made beer), schemes Dad has thought up to make a bit of money. So now we are walking up Moorfield Road to get the tram again, only this time to go and watch Dad play in his band

somewhere in Wortley, and our carefree family of four has been joined by a fifth, a huge and threatening cuckoo, the double bass.

Knowing what is to happen, the family make no attempt to go upstairs, but scuttle inside while Dad begins to negotiate with the conductor. The conductor spends a lot of his time in the little cubby hole under the winding metal stairs. There's often a radiator here that he perches on, and it's also where hangs the bell pull, in those days untouchable by passengers, though it's often no fancier than a knitted leather thong. In his cubby hole the conductor keeps his tin box with his spare tickets and other impedimenta which at the end of the journey he will carry down to the other

end of the tram. The niche that protects the conductor from the passengers is also just about big enough to protect the double bass, but when Dad suggests this there is invariably an argument, which he never wins, the clincher generally coming when the conductor points out that strictly speaking 'that thing' isn't allowed on the tram at all.

So while we sit inside and pretend he isn't with us, Dad stands on the platform grasping the bass by the neck as if he's about to give a solo. He gets in the way of the conductor, he gets in the way of the people getting on and getting off; always a mild man, it must have been more embarrassing for him that it ever was for us. Happily this dance band phase, like the fretwork and the herb

Below left and below Looking down from Lower Briggate in June 1958. An ex-LNER 'B16' locomotive animates the scene on top of the bridge awaiting the 'right away' from City Station. Many people think that colour lights are a modern form of signalling on the railway, but this disproves that theory. The time, according to Peter Tong's toy shop clock, is 6.15 pm, as people board 'Horsfield' No 189 on their way home (189 is actually 180, having been renumbered temporarily). Although 180 was later to be preserved, it is more than likely that the steam locomotive ultimately became razor blades! *A. K. Terry*

Surprisingly little has changed in this city centre scene 34 years on. The Viaduct pub on the right remains, and although the shops have changed occupancy several times, the buildings themselves still stand, although in a much smartened state. Electrification is now a feature on the newly painted railway bridge with a great ugly bank of colour light signals dominating the view. *F. Whiley*

This is the service terminus outside Leeds United Football Club ground in Elland Road on 18 June 1955, just seven days before the trams would be replaced by buses. 'Horsfield' cars Nos 208 and 156 pass each other - 208 is returning to Meanwood, incorrectly showing the pre-war 23 service number instead of the usual 6. Apart from the Guinness and Mackeson adverts, there is also one for Anglo XL chewing gum - I can't remember seeing that for a long time! *A. K. Terry*

The same scene photographed on 1 September 1992. Elland Road is now so busy that it would be dicing with death to stand in the middle of the road in the same place as the shot taken in 1955. The whole scene has virtually disappeared. All that is left is a small piece of the football ground wall on the left-hand side next to the Ford advert. Even this will disappear shortly when the mighty new stand, as befits the last old-style First Division champions, is unveiled. *F. Whiley*

down Headingley Lane on a fine evening lifts the heart at the time just as it does in memory. I go to school by tram, the fare a halfpenny from St Chad's to the Ring Road. A group of us at the Modern School scorn school dinners and come home for lunch, catching the tram from another terminus at West Park. We are all keen on music and go every Saturday to hear the Yorkshire Symphony Orchestra in the Town Hall, and it is on a tram at West Park that another sixth-former, 'Fanny' Fielder, sings to me the opening bars of Brahms's Second Piano Concerto, which I've never heard and which the YSO is playing the coming Saturday. Trams come into that too, because after the concert many of the musicians go home by tram (though none with a double bass), sitting there, rather shabby and ordinary and often with tab ends in their mouths, worlds away from the Delius, Walton and Vaughan Williams which they have been playing. It's a first lesson for me that art doesn't have much to do with appearances, and that ordinary middle-aged men in raincoats can be instruments of the sublime.

Odd details about trams come back to me now, like the slatted platforms, brown with dust, that are slung underneath either end, like some urban cow-catcher; or the little niche in the glass of the window on the seat facing the top of the stairs so that you could slide it open and hang out; and how convivial trams were, the seats reversible so that if you chose you could make up a four whenever you wanted.

How they work is always a mystery. As a child I have difficulty in understanding that the turning motion the driver makes with the handle is what drives the tram, seeming more like mixing than driving. And then there is the imposing demeanour of the ticket inspectors, invested with a spurious grandeur on a par with the one-armed man who shows you to your seat in Schofield's Cafe, or the manager of the Cottage Road cinema in his dinner jacket, or gents' outfitters in general.

I don't recall anyone ever collecting tram numbers, but the route numbers had a certain mystique, the even numbers slightly superior to the odd, which tended to belong to trams going to Gipton, Harehills, or Belle Isle, parts of Leeds where

beer, doesn't last long. He gets bored with the fretwork, the herb beer regularly explodes in the larder, and the double bass is eventually advertised in the Miscellaneous column of the *Evening Post* and we go back to sitting on the top deck again.

After the war we move to Far Headingley where Dad, having worked all his life for the Co-op, now has a shop of his own just below the tram sheds opposite St Chad's. We live over the shop so I sleep and wake to the sound of trams; trams getting up speed for the hill before Weetwood Lane, trams spinning down from West Park, trams shunted around in the sheds in the middle of the night, the scraping of wheels, the clanging of the bell.

It is not just the passage of time that makes me invest the trams of those days with such pleasure. To be on a tram sailing

I'd never ventured. And Kirkstall will always be 4, just as Lawnswood is 1.

Buses have never inspired the same affection, too comfortable and cushioned to have a moral dimension. Trams were bare and bony, transport reduced to its basic elements, and they had a song to sing, which buses never did. I

Proof, if proof were needed, that Leeds was not all back-to-back housing and industry. It is 9 July 1955 at the terminus at Moortown Corner, where cars on the two Circular routes (No 2 via Chapeltown and No 3 via Harehills) would meet and 'lay over' for a while before both of them returned to the city. 'Horsfield' car No 169 is working a No 1 service back to Lawnswood, while 'Chamberlain' car No 124, freshly painted in red and cream, has just reversed, having arrived via Chapeltown as a 'straight' No 2 to Moortown before returning to Dewsbury Road on service No 9. 'Tizer the Appetizer' was a frequent advert seen on the front dash end of trams, and of course is still as popular today. *A. K. Terry*

Street Lane, Moortown Corner, on 1 September 1992. Unlike many parts of Leeds, this scene has remained fairly timeless - even the old-style telephone box stands in the same place. However, there are at least 13 motor cars in this modern scene. How many can you spot in the original? *F. Whiley*

was away at university when they started to phase them out, Leeds as always in too much of a hurry to get to the future, and so doing the wrong thing. I knew at the time that it was a mistake, just as Beeching was a mistake, and that life was starting to get nastier. If trams ever come back, though, they should come back not as curiosities nor, God help us, as part of the heritage, but as a cheap and sensible way of getting from point A to point B, and with a bit of poetry thrown in. ●

North Street, Leeds, in 1958. Note the small boys on the right in shorts, braces and 'pumps'. The hoarding on the wall on the extreme left is advertising Premium Bonds which, when this photograph was taken, had been on sale for just over 18 months. The jackpot prize then was £1,000 - quite a sum. The tram, 'Horsfield' No 178, has just passed Grafton Street and is followed by a Bedford Duple coach, fine vehicles which typify the period and were a common sight then; quite a few survive today as preserved vehicles. *A. K. Terry*

North Street in 1992. The left background remains, but the right-hand side has been long demolished, the trees and banking hiding new development and the inevitable high-rise flats. Few pedestrians use North Street today. It is a very busy one-way thoroughfare down to the even busier Sheepscar interchange. *F. Whiley*

No 2: Waxed bread wrappers

When I was at school my friends and I used to cover our textbooks with waxed bread wrappers. This was not only very practical in that the paper was strong enough to protect our books from most of the abuse they came in for during a typical school day, but it also looked pretty good, too - or so we all thought at the time.

When mass-produced sliced bread first came on to the market in 1925, wax-paper was the only suitable material available at the time in which to wrap it. However, sales of sliced bread did not in fact really boom until over 30 years later, when supermarkets were beginning to spring up all over the country and much bread was wrapped in plastic.

Bread wrapped in wax-paper brings to mind a more genteel era when shopping for groceries was an altogether more enjoyable exercise than it is today, a time of personal service, when every shop had a chair for waiting customers, and shopkeepers always wore a shirt and tie beneath their overalls. I'm not convinced that there was ever a time when ladies wore white gloves while out shopping for food, as suggested in our 1950s publicity photograph, but it's a nice thought.

There is good news for those with nostalgic yearnings. Because of its association with quality and tradition, wax-paper wrapping is on the way back, not only in the North where it is historically a favourite, but also in the rest of the country. ●

Julia Thorley

Photo courtesy of Warburton's

Large change

Today your High Street bank is almost unrecognisable from its 1950s ancestors. Pamela Howarth opens an account. . .

I was a young graduate transferring from a weekly to a monthly salary when I first opened a bank account back in the 1950s. Until then any money I had that wasn't in my purse found its way into a Post Office Savings Account. This worked very well except that in the '50s the most you could withdraw at any one time was £3, and sensible old Auntie Post Office was far too shrewd to let her clients loose with cheque-books. There was nothing to stop me paying my monthly cheque into the Post Office, but psychologically I needed the status of a bank account.

Status was the name of the game in those days. Not every Tom, Dick or Harry could open a bank account and, far from wooing young customers, banks kept them at arm's length, demanding character references before they would deign to accept them as clients.

I didn't so much shop around for a bank as look and listen. My father banked with Lloyds, so naturally I wanted somewhere different. It was like picking a racehorse because you liked its name. National Provincial and Midland sounded too humdrum; Barclays would have had more going for it if it had been spelled the aristocratic English way, Berkeleys, instead of the plainer Scots version; but Westminster had the right ring, so I plumped for that. Little did I or anyone else dream that in 1970 it would marry beneath itself and become plain NatWest.

I remember bowling into the branch at the corner of Euston Road and Hampstead Road one sunny lunch-hour and asking if I might see the manager.

'Yes, madam,' replied the teller, and two minutes later I was facing a fatherly figure across a broad, heavy desk ready to entrust my entire financial resources to his stewardship. By today's standards he played hard to get, wanting to know how

> *Banks had dignity with polished counters and polished panelling round the walls to match.*

long I'd been in my job and at my address, as well as the names of two character referees; but after we'd shaken hands and I was once more outside on the pavement I suddenly felt ten feet tall.

The post worked like a charm in those days, and the next morning my more worldly friend received the reference request. 'What does it matter whether you're respectable or not?' she demanded indignantly. 'They're getting your money, aren't they?'

But I was not to be put off. As far as I was concerned there was a certain cachet about holding a bank account. It was another milestone along my path to adulthood, another qualification and sign

that socially I'd arrived. As for getting my money, I'd no idea how banks worked. For all I knew my bit might have been labelled and tucked away in a strongroom doing nothing.

Sadly that particular branch was a casualty of a road-widening scheme. Recently, when I asked if its poor old vaults had been filled with cement before the tarmac was laid, I was told that they were probably incorporated into the subway system.

Once you'd jumped through the prescribed hoops to gain admittance, banks treated you very well indeed. For years I continued to find the manager as accessible as on the day of my first visit. Only recently did I discover that nowadays you are required to make an appointment.

In the '50s the tone was formal and courteous, every transaction beginning with 'Good morning' or 'Good afternoon', which is friendlier to my ears than a casual 'Hello' and the casual attitude that sometimes goes with it.

In those pre-chequecard days, withdrawing cash from a branch where you were unknown presented no problem at all. Someone simply hopped behind the scenes and rang up your branch. Long-distance calls were chargeable, but otherwise the service was free.

Banks had dignity with polished counters and polished panelling round the walls to match. Here the client found everything he needed - pens standing in their own little inkwells, blotters, paying-

This 1950s Westminster Bank interior is plain and functional - severe, even. What little decoration there is matches the reproduction Queen Anne style of its exterior. Note the tiled floor, lack of electronic gadgetry and, of course, the absence of elaborate security screens.

The photograph on the right shows how the same branch was modernised - ceramic tiles have been replaced by vinyl tiles, the ceiling is lower with more diffused lighting, and the far wall has been jazzily wall-papered. Still a little cold-looking, though.

The modern bank interior (not the same branch) is airy and spacious, and the atmosphere much more informal than its 1950s counterpart. Customers may help themselves to the free literature available on a huge range of financial services - and there are now 'self-service' points at which they can withdraw or pay in money, check on their account balance or request a statement or new chequebook. *All National Westminster Bank PLC*

in slips, ashtrays, and elegant little roll calendars in silver holders. Sometimes the ink was rather dubious and the pen nibs scratchy, but the thought was there. As for the calendars, they seemed to survive amazingly well, and if there was evidence that a young child had tampered with the date, another client would do the decent thing and put it right again.

Although there was nothing to prevent a woman from becoming a teller, in the '50s the majority were men; but in most branches you could see women in the background operating National Cash Registers.

That marvellous modern institution, the central queue, filtering customers to tellers as they become free, had not been dreamed of. Instead there were queues at every position and it was something of a gamble which one to join. I soon became bankwise enough to recognise that a short queue was not necessarily a fast one. It was quicker by far to stand behind a line of people waiting cheque in hand to draw cash than one man with a bag or attaché-case lurking at his feet. The odds were that when his turn came he'd heave it up to the counter and pay in bag after bag of coins and bundles of notes and cheques.

If time was no object it was fascinating to watch coins being shovelled on to scales and weighed. At least it was quicker than watching the teller actually counting the coin of the realm. First the silver, or cupro-nickel: eight half-crowns to the £1, 10 florins, 20 shillings, 40 sixpences or 80 threepenny bits. The latter co-existed with and were heavily outnumbered by large, brassy, 12-sided threepenny bits, but it was the silver ones that people coveted to put into Christmas puddings for luck; anyone finding one in his helping could make a wish. Then there were the coppers or bronze - 240 pennies to the £1, 480 ha'pennies or 960 farthings.

Although banks accepted and dispensed the last two, they would have no truck with anything like vulgar fractions on cheques. In shops you had to round the ha'penny up or down and either tender an odd one or take one in change.

With the 1993 withdrawal of the large 10p coin, originally the florin, the last of this ancient coinage has disappeared from our purses and pockets forever. Our notes have suffered a similar fate. In the '50s there were large, white, crisp fivers and notes for £1 and 10 shillings, commonly known as 10 bob. Nice as it was to have the occasional fiver pass through your hands, they were sometimes difficult to change in small shops and coffee bars.

One of my boyfriends kept a 10 shilling note nest-egg in his breast pocket for emergencies. At a pinch it would keep him in cigarettes, lunch and fares to work for a day, or cover two tickets to a cartoon cinema in the evening, followed by two small snacks and the fares home.

'One day,' he used to promise me, 'that 10 bob'll be a fiver.'

I was duly impressed by such ambition; but the days of looking on a fiver as a respectable sum have long departed, and I wonder what denomination of note he carries in his breast-pocket today.

The female version of the nest-egg appeared on charm bracelets in the shape of a minute, glass-fronted box into which a £1 or 10 shilling note had been cunningly folded.

My first chequebook filled me with pride and to this day I rarely leave home without one. You can keep plastic money - as far as I'm concerned it just doesn't feel as good. Armed with that chequebook, for the first time I was free to buy at a moment's notice, instead of spotting the dress of my dreams in Swan & Edgar - now sadly no more - and then having to go to a Post Office to withdraw cash and return to the shop to make the purchase.

Marks & Spencer was practically all counters in those days, and an airy 'Will a cheque be all right?' created quite a stir. For starters, a supervisor had to be roped in to oversee the transaction and a writing surface of sorts had to be improvised. Often the assistant would obligingly fold back a pile of jumpers to provide a suitable space. Identification presented no problem. The classiest thing to produce was a driving licence, but as there were fewer of these around than chequebooks, stores were satisfied if you could produce a letter addressed to yourself that had been through the post, as long as it tallied with the name and address you'd carefully printed on the reverse of your cheque.

In time there came a day of reckoning with my first bank

They always make me welcome

And why not? We do not regard an account as being of small consequence simply because it *is* small. On the contrary, we believe that the importance of any banking transaction lies in its importance to the customer. We try to meet our customers' requirements in that spirit and to give the same welcome and the same friendly attention to all of them, whatever the size of their accounts. If you would like to know more about the personal quality of Westminster Bank service, the Manager of your local branch will be glad to tell you

WESTMINSTER BANK LIMITED

In these 1950 advertisements for Westminster Bank, as it was then, it sets out to attract both men and women of good character, promising friendly attention and personal consideration to all its customers. It would welcome 'the craftsman and the housewife no

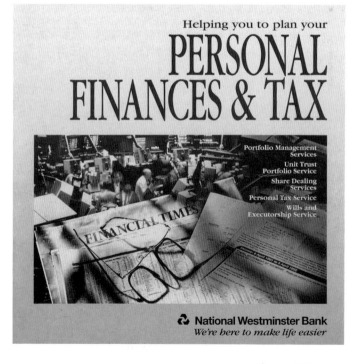

Helping you to plan your

PERSONAL FINANCES & TAX

Portfolio Management Services
Unit Trust Portfolio Service
Share Dealing Services
Personal Tax Service
Wills and Executorship Service

National Westminster Bank
We're here to make life easier

All sorts and conditions of men

Amongst people of all classes and every kind of occupation, 'banking with the Westminster' is becoming more and more a normal feature of everyday life. The Westminster Bank gives the same friendly attention and personal consideration to all its customers, without regard to occupation or income; it welcomes the craftsman and the housewife no less than the professional man and the business woman. If you would like to make use of this personal banking service the Manager of your local branch will be glad to help you.

WESTMINSTER BANK LIMITED

less than the professional man and the business woman'.

Today National Westminster Bank claims that 'We're here to make life easier', whether you need help in running a small business, or advice on portfolio management. *National Westminster Bank PLC*

statement. In the '50s you knew at a glance if you'd overdrawn because it was printed in red. There were also deductions for bank charges and 5 shillings for chequebooks, both of which I was happy to pay for the privilege of holding an account. Pocket calculators hadn't been invented, so if you wanted to check the bank's arithmetic, you had to do it the hard way.

Gradually I was to learn of the bank's other services, like standing orders, foreign exchange, deposit accounts and loans, but the range then was small compared with today's.

Two things were to revolutionise banking in the 1970s - decimalisation and computerisation. After that, plastic money really took off.

Any latterday Rip Van Winkle who dropped off in the 1950s for 40 years rather than 40 winks would be in for a series of surprises if he woke up in a NatWest bank today. Not only are coinage and notes entirely different, but interior decor has also been modernised beyond recognition, giving a more spacious, open look. In most branches there are whole racks of booklets explaining the services now of offer, such as *Starting your Business*, *Running your Business*, *Personal Finances & Tax*, *Mortgage Services*, *Pensions & Life Assurance* and *Insurance Protection* to name but a few. Credit cards are dealt with in *Business Card* and *Credit & Payment Cards*, which cover everything from NatWest servicecard, cashcard and chequecard to the Switch facility and the up-market Gold Plus service. Tucked away somewhere there is also usually a department dealing with stocks and shares.

If, like policemen, bank customers seem to be getting younger, it's because they are. Catch 'em young is the name of the game nowadays with facilities for students unheard of back in the '50s. NatWest now invites 18-year-olds and over to apply for credit cards and provides a free helpline for those requiring assistance. Rip Van Winkle would feel *really* old. . . ●

MONETARY MILESTONES

1960 By the mid-1950s, inflation had rendered the farthing practically worthless. Production ceased after 1956 and they were withdrawn at the end of 1960.

1965 The decision was taken that Britain would change to a decimal system of coinage in 1971.

1968 The 10 new pence and 5 new pence (equivalent to the former florin/2 shillings and 1 shilling) were introduced.

1969 (1 August) The old halfpenny was withdrawn.

(October) The seven-sided 50 new pence (10 shillings) coin was issued.

1970 (1 January) The old half-crown (2s 6d) was withdrawn.

(21 November) The 10 shilling note was withdrawn.

1971 (15 February) The official changeover to decimal currency took place.

1982 The 'new' was dropped from cupro-nickel coins.

1983 (21 April) The pound coin came into circulation.

1984 (December) The new halfpenny was withdrawn.

1988 (11 March) The £1 note was withdrawn.

Piccadilly phones

Piccadilly Circus Underground station was completely rebuilt in the 1920s. The work, costing half a million pounds, was begun in 1924, and despite the fact that the huge circular concourse was entirely underground, was carried out with minimal disruption to surface traffic. Set into the circular wall were the stairways to the various street exits, basement shop windows for Swan & Edgar's department store, a booking office, and this bank of telephone booths, photographed in 1949.

The booths, as can be seen, were finished in bronze with a frieze of Tivoli marble. The columns supporting the ceiling were of imitation stone with narrow brass fillets at the angles and twin lampshades at the top - as can be seen in the 1992 photo, these still exist.

Admittedly the 1949 picture appears to have been taken in winter and the more recent one in high summer, so the contrast between the hats, overcoats and fur coat and the short skirts, casual trousers and shirt-sleeves is especially accentuated.

Perhaps the greatest contrast, however, is between the privacy of the heavy wooden sliding-door booths and the

'open-plan' layout of the 1990s, without even a sound-proof hood to insulate you from the conversation next door. The loss of the woodwork also, of course, enables the provision of at least twice as many phones.

The list of exchanges, both national and international, on the wall in 1992 reminds us that in 1949 calls were charged by distance, rather than time, and had to be routed through the operator. In 1958 Subscriber Trunk Dialling (STD) was introduced, making it possible to dial direct, eventually virtually anywhere in the world. ●

1949 photo BT Archives © British Telecommunications plc 1992; 1992 photo Will Adams

The Martyred St Ann

Keith Howard reflects on the transformation of St Ann's Road, Harrow

St Ann's Rd, Harrow - two decades saw the old individual shops replaced by a vast precinct, looming up like a cliff face of geometrical perfection, but lacking the former soul and character.

St Ann's Road, Harrow, in the 1950s was akin to Mother Earth in one very important aspect; you could live out your entire life within its precincts without ever having to leave it. Along each side of that few hundred yards was virtually every shop you could possibly need - plus a few that you wouldn't, but even they had curiosity value.

The familiar glazed tile frontage of the United Dairies smiled a greeting from its vermilion and white fascia, welcoming you into its cool, marbled parlour of farmyard comestibles, now frowned upon by the health fanatics. Further along was W. H. Smith's, but not the gleaming, computerised cavern in which we can lose ourselves today. This one was more of a dark and cosy grotto wherein could be found helpful denizens lurking among the newspaper-buttressed, magazine-festooned recesses. Pliable plastic records of nursery rhymes could be purchased for the children - opaque discs, all colours of the rainbow.

Dignified industry thrived next door where Axon's the Ironmongers (later to become John Dyas) opened its doors to distressed DIY victims. Earnest and serious men in grey warehouse coats dispensed mysterious domestic whatnots from innumerable little brass-handled wooden drawers behind the sawdust-scented counters. No customer was left floundering helplessly among bar-coded plastic packets. There was no computer check-out looming up like a robot sentry. Instead, ornate tills with melodious bells reckoned up the passing of half-crowns, florins and their smaller

much-loved and missed brothers.

Each night these takings, along with those from the neighbouring shops, would be banked in the night safe of the National Westminster Bank on the opposite corner. Happily this building has survived in its entirety, a magnificent structure of dark red brick, sufficiently small-scaled to form the splendid columns that flank its sides. In recent years the interior has been gutted of its original mahogany fittings and the white-collared staff are but a memory, but no trace of these internal transformations has been allowed to mar its grand exterior. Like a smaller sun-tanned version of Buckingham Palace, the NatWest has survived.

Nestling in a neat row in the shadow of the bank were St Ann's Cottages, snug-looking terraced homes with long front gardens, and beyond them several smaller shops such as Bon Ton's the sweetshop, smelling as all sweetshops did in those days of that magical mixture of chocolate and fragrant pipe tobacco. At Christmas the windows would be arrayed with those mysteriously exciting parcels of cotton wool, trimmed with festive tinsel of red and green, containing within their marshmallow interiors all manner of seasonal novelties to brighten the infant eye. Conveniently enough, fairy lights could be purchased from the electrical shop next door, along with an assortment of light bulbs, all colours of the rainbow - even red, white and blue striped!

On the corner where Havelock Place meandered into the back of the Havelock public house was St Ann's very own *boutique fantastique*, Lafoy's Music Shop. How we would gaze in awe at the dusty piles of sheet music, those richly coloured musical instruments fashioned like ceramic potatoes; the tin kazoos like miniature submarines with their metal-meshed conning towers; mouth-organs guaranteed to drive parents mad within minutes and bring sores to the corners of lips within a day. Black, purple, silver and gold dominoes stared back at us, blind-eyed and enigmatic, while rubicund canvas faces of friendly clowns and frightening skulls

mocked us silently, defying us to out-stare them. There was magic and enchantment in that shop if only we had pennies enough to spend.

On the opposite corner of Havelock Place stood Greenhill School, a single-storey building but nonetheless a school - you could never mistake that stolid citadel of learning for anything else, with its distinctive, small, square-paned windows. How different from the glass and cement cubes that have taken its place in these advanced times.

It goes without saying that food shops of all descriptions abounded. A real fishmonger's with real wet fish; a real butcher's, baker's, grocer's and at least two greengrocers' - all with real people working in them who had the time to ask after your health and when you were going on your holidays. Oh yes, the inner man was well catered for, with even yet another public house at the far end of the road.

And the 'outer' man and woman was catered for, too, for there was a

> *... shops of all descriptions abounded. A real fishmonger's with real wet fish; a real butcher's, baker's, grocer's ...*

shoeshop as well as Norman's the dress-shop, at the corner of Angel Road, with its sedate but stylish dresses, costumes and coats. And even if you didn't see what you wanted in the window they were sure to have it inside.

For gentlemen of all ages Brayley's (later to become Fosters) would clothe all from eight to 80-plus. How well I remember being taken there, either for a new 'Sunday best' suit or something more substantial for school. The assistants - elderly to the infant eye - in their three-piece bird's-eye suits, wearing their tape measures around their necks like mayoral chains of office, gazed down with world-weary eyes and exuding an air

of sardonic benevolence: 'I think you will find that quite suitable for the little gentleman, madam. And it allows room for growth, too', while doubtless thinking 'The little wretch'll have the seat out of those in a fortnight'.

But after the dull chore of trying and buying school clothes there was the excitement of the nearby toyshop - The Dolls' Hospital, where the famous Pelham Puppets were strung up on parade along the back of the window like the colourful cast line-up of a Christmas pantomime finale, frozen technicolour smiles beaming, strung limbs held forever in some weird arabesque. And before them were the tiers of miniature Sunbeam Talbots, Chad Valley clockwork delivery-vans and Lilliputian Trooping of the Colour figures in their black bearskins and gleaming red tunics at sixpence each. And because it was called The Dolls' Hospital, many a beloved but broken Mary Jane with a fractured skull or a detached limb could be made better, and a much-loved teddy bear would have a button eye replaced or a squeak restored.

Further down and opposite, amid the selection of second-hand shops, could be found Philip Lewis, Tailors. You knew he was good because he never deigned to put anything in his windows. How could he, for being a bespoke tailors his wares were yet to be made.

The Greenhill Laundry proudly proclaimed itself with a colourful wooden banner above its entrance, decked out with a beautiful painting of Harrow on the Hill. This was to become Pimm's Popcorn Factory which had its own charm as it exuded the delightful aroma of custard the length of St Ann's Road.

Beyond the laundry an imposing block of tall houses reared up to the sky, aloof behind the stately railings that bordered their shrubberied front gardens. No doubt moneyed people dwelt there once, borne through domestic life by the efforts of faceless cooks, subservient housemaids and ubiquitous 'tweenies. But these minions had long since departed due to the traumatic changes wrought on society by the Second World War.

Meyer's second-hand shop boasted trestle tables laden with boxes of fragile 78 rpm records, ranging from popular musicals to Richard Strauss's 'Death and Transfiguration'. Law and Field's Gents Outfitters on the corner of Byron Road was nothing ostentatious, but at least they sold the much-sought-after Meridian underwear for discerning gentlemen of mature years. Malins would repair your razor as well as repair your appearance with a short-back-and-sides trim. Then there was Endean's the Leather Shop. Endean's sold real leather goods, not some synthetic merchandise vat-born in an oriental laboratory. Their stock was robust and studded with brass. Not for them the kinky black leather of today; they purveyed the honest gleaming brown leather of horse brasses that the Yeomen of England have known and loved throughout the centuries.

We are nearly at the end of Memory Lane, but let us cheat a little and peep around the corner to what was Clarendon Road, the prettiest road of shops in Harrow, with its line of flourishing trees softening the concrete and brick. Here could be found a quaint little bookshop with its rows of dusty domes and the faces of long dead monarchs staring from the multi-coloured rectangles of foreign stamps; Spivak the jewellers, with its window full of treasure trove; and the Norfolk House Butchers who sold the flakiest, tastiest boiled bacon in town.

Opposite was Somerton's, the ladies' dress shop, with its elegant wrought-iron stands and its elegant wrought-iron lady assistants. With the hauteur of ice maidens, these chic, willowy high priestesses of fashion could be seen titivating incredible creations of millinery on slender metal stands where they remained poised with gyroscopic impossibility.

Martin Green's, the respectable, no-nonsense Gents' Outfitters, was where you could purchase a Paisley or Churchillian spotted bow-tie that you actually tied yourself. Aged but learned seers of their trade attended to your needs from behind oaken and

This 1967 newspaper photograph, picturing the other end of the road, was captioned: '. . . Harrow's Saturday morning traffic grinds slowly along St Ann's Road at noon. Since the introduction of the one-way traffic system in Central Harrow this has become a common sight, occasionally further confused when some errant motorist turns left from Station Road into the main stream. A further restriction is on the way. Harrow Council has received permission to alter the stated time for "no waiting" from 8.30 am to 8 am in the light of experience of an earlier build-up of traffic than was first visualised.'

Today the busy junction with Station Road is a pedestrianised area. Like it or not, it's solved the traffic snarl-up problem!

glass counters that resembled sartorial pulpits.

But my favourite shop had to be Mancini's where you could find a matchless selection of home-made sweets. One window was devoted to tiers of little glass terraces, each with its own dainty pyramid of violet creams or virtually any other confection you could imagine or desire. The other window was a Disneyland rainbow playbox of toys. Many a child had to be dragged away screaming from that Aladdin's Cave by a long-suffering parent. I know, for I was dragged away every Saturday afternoon.

Now we return one last time to what was probably the landmark of St Ann's Road. Flanking the far end, a broad building of red brick, Adam's the Furnishers looked as if it would still be there at the Last Trump. To those who did not know, the strange upper storey with its open portals ranged along all four sides of the building must have presented a mystery. But this establishment had started life as one of the pioneers of the Harrow cinemas, the Picturedrome.

But one corner of the old Picturedrome building was given over to something thankfully rare; a noxious little shop where horsemeat could be procured, presumable for the feline members of the family. Even now I find myself holding my breath when passing the place where that shop used to stand. Would that Pimm's Popcorn Factory had been in existence then to soothe the olfactory organ with its vanilla balm.

Alas, all that has gone now. The Vandals and Visigoths of Property Development have swept down to the plains and razed to the ground the well-remembered shops in a welter of architectural rape and pillage. Clarendon Road no longer exists and is now nothing more than a dark and draughty void, a space left over between a gargantuan glass-flanked office block and a concrete helter-skelter which disgorges a never-ending stream of growling ego-boxes from the upper levels of a car park.

And what of St Ann's Road itself? Now each side is ranged by monoliths that block out the sunlight and turn the pedestrianised area between into a wind tunnel. The small food shops have vanished and are no more than ghosts of a cherished memory, and the only trace of the old world are the upper storeys of a row of houses above the aluminium and glass portals. These cottage windows with their decorated stone mouldings are all that is left to remind us of much happier days - and the martyrdom of Harrow's own St Ann. ●

Change is happening all around us, and nowhere more so than in the High Street. Has your local shopping centre been transformed - or is one of those that has resisted change?
If you have some interesting 'past and present' photographs and some nostalgic recollections, drop us a line at the address on page 65.

Working the land

40 years of change on a Borders farm recalled by Isabel Dickson

What changes in 40 years, and none more so than in farming. In the 1950s an average Border farm could support perhaps four to five families, that is five or six full-time male workers, with wives, children and friends helping at specially busy times. Today all the advances in farming methods mean that now the average Border farm supports only one family, ie one man and a seasonal helper. So although the actual physical work is easier, it is now a much lonelier job.

In 1953 tractors were starting to take the place of horses, especially the grey Ferguson, small, uncomplicated and easy to drive. It had a simple gear lever and clutch pedal, and an accelerator lever which was attached to the steering wheel. Its speed was not excessive, but it was a sturdy machine which was in use for many years. However, horse-power was still in use, as this photograph shows. This machine was for raking hay up into long rolls ready for stacking, and it needed somebody to sit up top and work the lever that raised and lowered the prongs on the revolving drum which drew the hay into a row. The horse moved forward all the time, the prongs were raised and lowered, and after a few trips back and forth across the field there were neat heaped rows of hay ready for a 'Tumbling Tam' to pull into 'coils' (pronounced 'kyles'). These were large heaps of hay, and from these the hay was built around a tripod of poles into a haystack. The outside of the stack was raked smooth so that rain would not be able to penetrate too far. One man and a horse with a hay-rake could complete a 20-acre field of cut hay ready for 'Tumbling Tam' in a day. Then, of course, a squad would be needed for making the stacks. . . Wives and friends were often pressed into service here.

Above This 1963 photograph shows the next stage of development - a tractor and mower to cut the hay. First it was 'wuffled', ie shaken up, to let the air in to dry it by machine, and then the baler was pulled by the tractor along the rows. This gathered up the hay, packed it into cuboid shapes and tied them with twine, then ejected the bales on to a trailer which was being pulled behind the baler. When there were about eight to ten bales on the trailer, the driver pulled a cord, and the trailer tipped the bales in a heap on the ground. This made it an easier job to build them up in a block which would be weatherproof.

Above Today some grass is still cut and baled when dry as shown here. However, some is baled as soon as it is cut and each huge 'Swiss roll' is then wrapped in black polythene, using the machine in the fourth photograph, so that it is completely airtight. This will then keep all winter as silage for cattle, etc.

Right The new type of baler is also used for baling straw after the combine harvester has done its job, but this is not wrapped in polythene. Instead it is string or net wrapped and is used for winter bedding for animals. One man using these new machines can deal with five acres an hour. ●

Past and Present

Smile please!

Old boy Mike Whitaker goes back to school to see how today's pupils are shaping up

*I*n the mid-1940s, milk bottle tops were circles of cardboard, in the middle of which was a smaller circle which you pushed out, so as to insert a straw. Squirting milk all over myself was one of my earliest memories of St Petroc's Preparatory School at Bude in north Cornwall. I was but a day-boy - one of a handful - and my membership of the school community was therefore pretty tenuous. But, in my grey shirt, dreadful long grey shorts held up by an S-buckle belt, my cherry red cap perched on my wing-nut ears and my cherry red blazer firmly done up - all three buttons - a member I was. Which meant that I had to steer a carefully considered course on the half-mile walk to school.

Bude still contained a population of displaced Londoners, evacuees, coarse and brutal children, immediately identifiable by their uniform of long green cor-

duroy shorts and clumsy black boots, like scaled down Army boots. The evacuees did not get on with the native children, and neither group had much time for we St Petroc's boys. We were rather like exotic ducklings on a small pond full of predators. And we few day-boys belonged absolutely nowhere. But the fact remains that I, from the age of five, was able to walk to and from school, alone and unescorted, four times a day with no risk beyond having my nice cherry red cap whipped from my head and cast into a muddy brook (a not infrequent event).

School was spartan - the boarders had to enjoy cold baths - and all of us were walked in crocodile every day, rain, hail or sleet, either to the beach or through the lanes. The discipline was rigid. Canings were frequent, whether you had committed a crime or merely been present when one was committed. In this

life, I was told as I winced, the innocent must suffer with the guilty. And suffer we did. The teaching was sound and Public School orientated; I was studying French at seven and Latin at eight. There was a heavy emphasis on manners and behaviour - and no crime was greater than a lapse in manners.

Forty-odd years later I went back to the school. It was still in the same building, albeit extended, and inside the layout was exactly as I recalled it. Classrooms, changing-rooms, hall and staff-room were in the same place. The children still sat at desks with lift-up lids and inset inkwells. But there were differences, as a glance at the two accompanying photographs will reveal. The school now admitted - *girls*! Girls today form about half of the school's population. I was 11 and at the co-educational grammar school before I first encountered one of these strange creatures.

The uniform has changed a little. The startling cherry red of my day has mutated into a more sober dark red suit. But the atmosphere has altered beyond recognition. I was cowed by school; the children I saw were certainly not cowed. They were bright, lively and outgoing. In part, this reflects the fact that children in our contemporary society are allowed a voice and a view, whereas I had to be seen and not heard. Part is also due to a more relaxed regime - note that the staff appear in the current school photo, unthinkable 40 years ago. But some of it is due to the broadening of the school's mix; not only does St Petroc's admit girls, but also Service children whose background is not always the comfortable middle class one in which private education is traditional.

Overall I was surprised how little the school had changed. The cold baths have gone, but they still do Latin, they still have crocodile walks through the lanes, and the relationship between the school's children and town children is not much better. I was delighted to note, however, that the emphasis on manners remains - the children still stand up when an adult enters the classroom.

Those changes I did notice were all beneficial - the open and relaxed atmosphere, the broader social mix. I wish I could say the same for the school's environment. I recall the playing field being surrounded by open farmland. Now it is hemmed in by a spreading retirement bungalurbia.

I, by the way, am the miserable little scrap of humanity sitting down cross-legged on the extreme left of the front row in the *circa* 1948 school photo. ●

FOOD FOR THOUGHT

What do we eat more or less of than we did during and just after the Second World War? The following table shows household consumption of principal foods, in ounces per person per week:

	1942-6	1985-9
Meat	14	$12^3/_4$
Fish	$7^1/_4$	5
Bread	$60^1/_2$	$30^1/_2$
Milk (pints)	4	4
Butter	2	$2^1/_4$
Eggs (no)	$2^1/_4$	$2^3/_4$
Fruit	$12^1/_2$	$30^3/_4$
Potatoes	$67^3/_4$	38
Other vegetables	36	$46^1/_4$
Sugar	$8^3/_4$	$7^1/_2$

Taken from a study commissioned by Haliborange looking at the changes in Britain's diet and eating habits since the 1930s.

TALKING ABOUT THE PAST. . .

'Nothing is more responsible for the good old days than a bad memory.'
American humorist Franklin B. Adams

'Mind you, six bob *was* six bob in them days. You could buy three-penny-worth of chips and still have change from sixpence.'
Alan Bennett, 'The Lonely Pursuit', BBC, 1966

'The past is lovely because the past is complete: it's all there.'
American novelist Gore Vidal

Barnardos

Julia Thorley describes the development of one man's vision across the decades

At the age of 17 Thomas Barnardo felt called by God to serve as a missionary in China. But on coming to London from Ireland he met the appalling conditions of thousands of destitute children living and dying on the city streets. So instead he channelled all his evangelistic zeal and energy into saving these 'street urchins', and established Dr Barnardo's Homes to fulfil his vision of providing a better life for them. That was in 1866. The charity he founded transformed the care of destitute children throughout the United Kingdom into a national concern, and by the time of his death in 1905 the organisation cared for over 62,000 children, and its annual income was well over £200,000. The work of the charity today costs £1.5 million a week.

When Barnardo's is mentioned today, most people automatically think of orphanages, but in fact the last traditional long-stay children's home was closed in the early 1980s. Today residential care is provided for more than 600 children and young people, but those unable to live in their own homes are cared for wherever possible in 'ordinary' houses in 'ordinary' streets. There are also schools and hostels for children who have emotional difficulties or handicaps where they can move towards independence.

The range of young people receiving help from Barnardo's has expanded. Where it was once primarily the homeless, the organisation now operates with the much wider brief of giving children, young people and families facing disadvantage or disability the best possible chance. This includes providing a stable home life and security, to get young people off to a good start; work with the Training Agency to provide youth train-

As part of its new modern identity launched in 1988, Barnardo's unveiled a new logo, used as the title of this feature. The old image (*below*) was two young children inside a protective circle, but the overall effect was similar to a 'no entry' road sign and suggested that the children were somehow excluded from the world around them. The new logo (*top*) shows three figures who may be male, female, young or old, users or carers, volunteers, staff or donors - in fact any form of partnership helping overcome the barriers to as full a life as possible. It conveys the positive and active nature of Barnardo's, its spirit of energy and drive. The figures are portrayed in green - fresh and natural, the colour of growth. It is also non-racist, and the international safety colour.

ing schemes to help young people fulfil their potential; Intermediate Treatment (IT) programmes which help young offenders or would-be offenders to develop a more positive attitude to life and provide an alternative to custody; advice and support for drug abusers; providing a lifeline for parents in difficult circumstances by offering advice and support via family centres; training for the disabled and those with severe learning difficulties; and projects for the 16-plus group who are either leaving care or who have been rejected by their family.

Youth and community work helps young people towards independence, diverts others from crime, and helps communities to help themselves. Family support is available for parents caring at home for a young person with a handicap, while there are day care centres for vulnerable families with children under five.

Since the charity was founded, each decade has brought its own changes within society, but in the 1960s Britain underwent particularly eventful social upheaval which forced Barnardo's to reassess its priorities. After the turmoil of the Second World War, Barnardo's had continued its sterling work, but had perhaps become a little complacent in its attitude. It had been concentrating its efforts in three main areas: bringing up children in its network of children's homes, placing them in foster homes (or, very often, a combination of these two), and operating a programme of support - both financial and with actual goods - to help families who wanted to keep their children with them, but were finding it difficult. However, the social changes in the 1960s showed up this system to be rather blinkered to what was happening in the real world.

The new 'pop' culture emerged, the divorce rate rose, and the arrival of drugs on the youth scene posed new temptations and threats. The Barnardo's systems failed to take account of the burgeoning Welfare State. Local authorities began to develop better and more imaginative projects to keep children at home. The old philosophy that the Barnardo's family was the best environment in which to grow up had overlooked the value of staying with one's own family. While in the 1990s Barnardo's is committed to the belief that every child thrives best in a family situation, in the 1960s there was a need for a new philosophy to reflect the changing social scene, and readjustment to meet new or newly emerging needs.

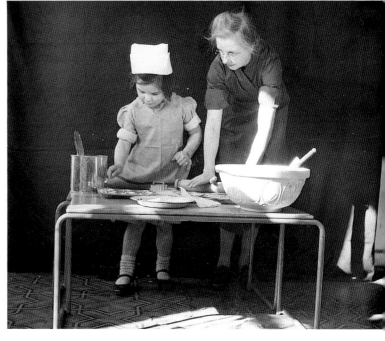

During the 1950s, young girls were still being educated for a life in domestic service. Even very young children were taught to cook, while also pictured here (*top*) is a group of boys being trained in a bakery.

The emphasis has now changed, and in the 1990s Barnardo's popular Caribbean restaurant in Leeds (*left*) provides full commercial catering training. *Barnardo's*

The placing of babies for adoption was one of Barnardo's main roles in post-war Britain, when in 1947 Barnardo's became a registered adoption society, building on its earlier extensive fostering work. There was great social pressure for unmarried teenage mothers to give up their children, and even had they wanted to keep them it is unlikely that they could have afforded to. However, in the 1960s the availability of contraception reduced the number of unwanted pregnancies, and there was less social stigma attached to being a

Training young people to take their place in society has always been an important function. During the 1950s these uniformed young boys were been trained for a career in gardening. In the 1990s, horticulture is still an area where youngsters are trained. Here a young wheelchair-user is being instructed in nursery skills. *Barnardo's*

The need to raise funds is ever-present. Today's appeals ads are much more hard-hitting than their 1950s equivalents. *Barnardo's*

DR. BARNARDO'S HOMES:

NATIONAL INCORPORATED ASSOCIATION
STEPNEY CAUSEWAY, LONDON, E.1

How you can help—

10/-	will help towards the cost of the children's food.
£500	For this sum a bed can be named for 20 years.
£1,000	For this sum a bed may be endowed in one of our Branch Homes, Hospitals or Training Schools. *A bronze plate will be affixed and particulars of the occupant sent to the donor from time to time.*

Will you make a gift in memory of some dear one?

YOU CAN ALSO HELP } By transferring Stocks and Shares to the Homes. By making your Will and leaving a Bequest to the Homes.

To Dr. BARNARDO'S HOMES
Stepney Causeway, London, E.1. .... 19........

Herewith I enclose a gift of £ : :

to help your NATIONAL WORK AMONG DESTITUTE CHILDREN.

Please acknowledge to me as below:

(Name)..
BLOCK LETTERS (Mrs., Miss, Esq., Rev. or Title)

(Permanent Address)..

....

....

Cheques and Orders payable " Dr. BARNARDO'S HOMES " and crossed
" BARCLAYS BANK LTD., a/c Dr. BARNARDO'S HOMES "

ANNUAL REPORT, 1952

TRACY HASN'T TALKED FOR 3 WEEKS. BUT SHE'S TOLD US ALL WE NEED TO KNOW.

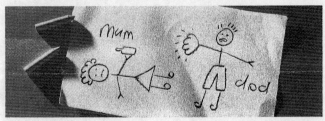

It was when Tracy came to us for day care that she began to play out her anxieties.

Her story proved to be an all too familiar one; a depressing tale of poverty, drink and violence.

So we involved Tracy and her family in one of our Family Centres. This not only gave her lonely mum a chance to meet new friends but, importantly, she and her husband were also able to attend counselling.

Because the whole family came into our centre, Tracy didn't have to go into care.

That's the way Barnardo's work. Help the family and you help the child. Sometimes, all that's asked of us is a sympathetic ear. But it could just as easily be practical guidance on housing or finance.

We work closely with families who are living with disability, or the emotional effects of child abuse. We also advise new parents on how to be good parents.

Our whole aim is to see children grow in a stable, loving family environment. However, our staff and budget are increasingly stretched. And if we're to continue, we need your support.

So if you would like a fuller picture, or to make a donation, please write to Barnardo's, Tanners Lane, Barkingside, Ilford, Essex IG6 1QG.

Barnardos
TOGETHER WE CAN GIVE YOUNG PEOPLE A CHANCE

single parent. Also, although it is by no means easy to bring up a child on state benefits, it should be possible, theoretically at least.

So as the number of babies available for adoption dropped, Barnardo's turned its attention to placing children who were previously thought to be unsuitable for adoption - older children, those who have been in care, or who are disabled, or in some other way traditionally thought of as 'hard to place', of all ages, ethnicity, ability and backgrounds. In the 1950s the legal framework was set up to separate natural and adoptive families. Today the whole process is much more open, and the two parties can work together for the benefit of the children.

From the 1970s onwards improved welfare benefits were intended to ensure that no family was ever put in the position of having to give up the care of its children for financial reasons, so Barnardo's began to close its traditional children's homes. Similarly, there were over 11,000 children under 16 in mental hospitals in the early '70s and a project began to bring them back into more normal domestic circumstances, with help and support.

Sadly, many of the problems that first prompted Dr Barnardo into action are still with us. In the 1990s Barnardo's has launched an 'Agenda for Action' targeting three major areas: homelessness, child sexual abuse and HIV/AIDS.

Homelessness is of course where Dr Barnardo started, yet the problem continues. Although precise figures are difficult to obtain, there does seem to be an increase in the number of homeless young. Every night the number of homeless people in Britain is, according to Shelter, 363,500, equivalent to the population of Coventry. Estrangement within families today is, broadly speaking, caused by family breakdown (one in three marriages ends in divorce), abuse - physical, mental or sexual - which creates a need to escape, or a situation in which parents and children just don't get on. Although homelessness was a problem in the 1950s and '60s, it was not as common as today, partly because in a time of high employment it was easier for young people to find work and so afford accommodation. Recent changes in benefits mean that today there is less of a safety net.

Sexual abuse is a newly emerging problem. The number of reported cases is increasing, but whether this is because the abuse has grown or because it is being brought out into the open is not known. However, in some Barnardo's projects more than half the clients - parents as well as children - have been victims of sexual abuse. Barnardo's was one of the first signatories of the Declaration of the Rights of People with HIV and AIDS to prevent discrimination, and the charity as a whole is very committed to its work in this area, not as experts on AIDS but as experts on children. It perceives a need for education of children from the age of 11, and there is a lot of work to be done in overcoming prejudice. This work has provoked sharp criticism in some

In the 1990s Barnardo's shops have lost their dowdy 'charity' image, not only in their outward appearance, but also in the goods they sell. *Barnardo's/Julia Thorley*

quarters, particularly the organisation's involvement with the Terence Higgins Trust in the production of educational leaflets for school children and parents.

The old paternalism of Dr Barnardo's Homes has gone, and when the charity changed its name to Barnardo's in 1988 it confirmed the organisation's move away from the outmoded Victorian image. The charity was founded on the inspiration and values derived from the Christian faith, but it has developed and is enriched and shared by many people of other faiths and philosophies. Today children are given more credit, more of a say in what happens to them, and there is now a statutory duty to take into account the wishes and views of young people. Barnardo's is poised to tackle the problems of the 1990s - sadly, it looks as though they, and other organisations like them, will always be needed. ●

Oxfordshire roads . . . past and present

The second volume in Past & Present Publishing's new roads series deals with Oxfordshire, a county rich in contrasts for the motorist. Sleepy villages that time has passed by (some of them only sleepy because recent bypasses mean that traffic also passes them by!), town centres struggling to cope with clogging traffic, and the coming of the A40, cutting a swathe through the north-east of the county and relieving pressure on many of Oxfordshire's former main trunk routes.

By way of an appetite-whetter, author **Susan Clark** presents three of the comparisons from the book.

> 'British Roads Past and Present: Oxfordshire' by Susan Clark is just published by Past & Present Publishing Ltd, and takes a nostalgic look at the county's highways and byways and the changes that have overtaken them in the last few decades.
>
> A large-format 96-page softback priced at only £10.99, it contains over 200 photographs featuring main roads and country lanes, town centres and market places, with special features on Abingdon and the MG, the Cowley Car Plant, The Highway Code, the RAC, and much more.
>
> On sale at your local bookshop, or available from the publisher at the address on page 65.

Left **THE COMING OF THE M40:** This is Waterstock, the starting point of the M40 extension. The first photograph shows the A40 as it looked in 1989, looking west towards Oxford. Two generations of roads are visible - the dual carriageway and, shadowing it on the right, the old London Road. At this time London Road was just a track leading to Holloway Farm, and ending in a curious 'B'-shaped section of road. Hedges mark the line of the disused Princes Risborough-Oxford railway across the top right-hand corner.

Right On 13 January 1991, three days before the M40 opened, this was the scene at the Waterstock intersection - to the other two, a third generation of roads has been added. The 'B'-shaped remnants of the old A40 and its successor, the A40 trunk road, are in the top left of the picture. But now, snaking between the farm buildings in empty six-lane splendour is the new motorway, veering northwards and crossed by three bridges. The nearest carries eastbound traffic from the A40(T) on to the motorway, the other two carry a footpath and the A418 over the motorway. *Photographs courtesy of The Director, Department of Transport, South East Construction Programme Division*

OXFORD HIGH STREET: This is the view from Longwall Street along the High, with Queen's College in the centre, as it looked in 1962. Traffic is passing freely, delivery vans can park and the cyclist has plenty of room. Two popular vehicles of the time can be seen - a Minivan is turning right into Queen's Lane and heading towards us is a 'bubble car'. These strange-looking runabouts were imported in the 'fifties, and several variations were manufactured by companies like Isetta, Messerschmidt, Daf and General Motors. This one appears to be a Heinkel three-wheeler, two-seater model. *Oxfordshire Photographic Archive*

In the 1990s the picture is quite different. It's almost impossible to appreciate the architectural beauty of the High, as it's so congested. Tourist buses cruise the streets: 'Every 15 minutes, every day' boasts the slogan. Cycle lanes have been created to encourage Oxford's cyclists and there is a pelican crossing. The Oxford Transport Study proposes to go further and close the High to through traffic during the daytime, except for buses and cycles. *Susan Clark*

SHIPTON UNDER WYCHWOOD: A 1949 scene in this delightful Oxfordshire location - outside the Shaven Crown Hotel is a solitary car, possibly a Vauxhall, while the postman on his rural cycle delivery somewhat self-consciously leans on his bike. He may be on his way to the village post office, then located in Church Street, by the war memorial in the distance. Stop lines marked on the road, like the one shown here, were first introduced in the 1930s. The direction sign would have been quite new, a replacement for the one removed during the Second World War. *Crown Copyright*

Any postman loitering on this spot in 1993 would be taking his life in his hands, as would any driver parking outside the hotel. The 'STOP' roadmarkings are now much more prominent. Though roadsigns were redesigned in the 1960s in the name of legibility, it is hard to see how anyone could think this untidy tangle an improvement on the earlier arrangement. It's interesting to note that the Shaven Crown Hotel's allegiance to the RAC seems to have lapsed, and that Ascott under Wychwood has moved a quarter of a mile further away. *Susan Clark* ●

Vintage cars

Remember your first Dinky or Matchbox car? Well, today those simple die-cast toys are collectors' items, as Joan Brittaine discovered.

It goes without saying that toy cars will be around so long as there are little boys to zoom them across the floor. What's more, the interest doesn't end there. When the same little boys become adults more than likely they will be looking around for toy motor-cars again - albeit for quite a different reason.

Serious toy collecting is now big business, in spite of being a comparatively new hobby. Twenty years ago toys were of little interest in the auction houses; now they fetch money which may be measured in hundreds of pounds - such as a Chitty Chitty Bang Bang car by Corgi recently on offer for £250!

For some people the lucrative side of collecting toys is the main driving force. But with the indisputable fact that most of us have a fascination for toys, with memories of our own treasures lasting well into adulthood, actually owning these toys is generally done for the sheer delight of it.

Nostalgia plays a big part in collecting - or it could just be a sense of achievement when a coveted toy as a child was out of the question. But whatever the motive may be, it is wise to assess the situation carefully in order to build up a meaningful collection rather than have a motley array.

Deciding on a theme is a good starting point - either a subject, a period in time, a certain scale of model or a particular manufacturer. And it's worth mentioning here that while looking around for toys of a bygone age, an eye should also be kept on the modern toys being turn out, which in their turn will be the future collectables.

Die-cast Dinky toys, or 'Modelled

Above One of the models that started it all: this Matchbox original issued in 1952 now fetches well over £100. Matchbox have now brought out another Coronation Coach which they see as a good investment for the future. *Phillips Fine Art Auctioneers*

Miniatures' as they were originally known, dominate many of the markets. But with their increasing popularity, people are becoming more selective - and it is wise to be prudent when choosing these for a collection.

Cars in mint condition and in the original box make the item more attractive and will certainly fetch higher bids; a boxed toy can add 20 per cent to the value in a sale. All the same, a word of warning would not be out of place. Handle and examine as many examples of die-cast toys as possible in markets or at swap-meets. Become familiar with them to such an extent that the genuine article is easily recognised. And this is not easy by any means, with repainting and faking so skilfully done nowadays as to be almost undetectable. It doesn't help either that replacement transfers and tyres are still available. So inspect carefully and as a guide it

Below The wheel turns full circle, and currently available is a series of 'Authentic Recreations' of the original Matchbox series produced 40 years ago. The 'Moko' trademark which appears on these models was adopted by 1948 when the first toy - the road-roller - was produced. In 1953, encouraged by the success of the miniature Coronation Coach pictured above, Lesney scaled down their larger toys to a miniature range. They were packed in a small box, the colour and shape of a matchbox - and the Matchbox series was born. *Matchbox*

is useful to remember that pre-war models usually had white tyres, whilst the post-war cars were generally black or grey. If the piece is in the original box, the buyer may be pretty sure of the real thing being inside. . . but a further word of warning; boxes are also being forged, so these should be closely examined as well!

Boxed sets of cars fetch the highest prices. Brisk bidding at Phillips, London, in 1992 for a No 4 Gift Set of Dinky Racing Cars with an estimate of £300 to £350 sent the hammer-price up to £520. But a record high must have been in the sale at Phillips in 1988 when a set of 22 Dinky toy vehicles, the first ever issued in 1934, sold for £2,200.

All these were forerunners of the immense variety of Dinky toys which were turned out - until competition arrived in the 1950s and '60s in the shape of Matchbox and Corgi toys. This brought about innovations and determined efforts at realism to win the attention of children and so increase sales. Plastic was used for the first time in features such as windows. Then came opening doors, boots and bonnets, and refinements such as jewelled front and rear lights and pseudo-chrome-plated parts appeared. It seemed there was no stopping the rapid development to make toy cars look like the real thing. Even popular television character cars were being produced, and interestingly these are already in the salerooms - like the Corgi set of a Batmobile and Batboat, a James Bond DB5 with wheel slashers, and a Lunar Bug, all boxed, which sold at Phillips in 1992 for £160.

However, there's no doubting the fickle nature of the toy industry. New inventions catch on quickly in a child's world and a toy that has sold well for many years can be wiped out almost overnight. Television provides the best opportunity for children to latch on to the latest craze, and this happened when friction toys emerged on the scene, with sales rocketing when children demanded cars which could be sent further distances with a single push. Then came the Transformers, which turned robots into cars by a few deft manoeuvres - and with serious discussion in the playground, the more complicated a transformer became the greater the attraction. These have not yet reached the salerooms, but it may be worth looking seriously at the more sophisticated models from a speculative viewpoint.

Dinky toys are making a comeback, and a 1991 model Chevrolet Bel Air is already changing hands for four times its original cost. And something else of interest by Matchbox is the Alarm cars which have their own ignition keys arming them against potential thieves!

Certainly electronic toys are holding the market at present. Microchip technology in the 1980s brought a range of cars with working lights, engine noise and re-directional techniques. And there can't be many children not owning a radio-controlled car. So with the swift development in toys nowadays it's anybody's guess what will be cornering the market in future auctions of collectables. It could be worth noting how long a toy remains popular - in some cases a short production run with few models being available places a toy high on the list of collectables. Generally items in sort supply mean longer queues and higher prices.

What is important is to buy the best you can afford - and it is far better to buy for pleasure rather than for the value. Build a collection that is interesting and attractive. Something to enjoy. This is what toy collecting is about. And who knows? Today's most attractive toys could be tomorrow's most attractive collectables. ●

In 1992, this Dinky 'Set No 2 Private Automobiles' comprising five American saloons in the original box was sold at auction for £1,980. *Phillips Fine Art Auctioneers*

Could these Matchbox Alarm cars be the collectables of the future? *Matchbox*

How many of you still treasure your childhood toys?
Drop us a line - with a photo if possible!

Lens of Sutton

45 years a transport literature specialist: Allan Mott meets the man behind Lens

To many a transport book reader, the town of Sutton, Surrey, is synonymous with one shop, Lens of Sutton.

When I first came upon this shop it was perched on the side of an old chalk quarry in Carshalton Road near the police station and not far from the town centre. To enter the shop was like entering another world, for it housed a unique collection of transport magazines dating back to their first publication date. When the site was developed the shop moved to 4 Westmead Road, still in the borough of Sutton.

For anyone who collects old transport magazines, Lens is always the first port of call in the hunt for a missing edition or two. The shop was established in 1929 by the original 'Len', railway enthusiast Mr T. Len. It was taken over in 1948 by its present proprietor John Smith, when he came out of the army.

Forty-five years later he continues to provide a remarkable service to the resident and overseas enthusiast, for not only does he hold old and new magazines, but he also has a huge photographic collection containing many views of old stations, locomotives and buses. ●

John Smith, proprietor of Lens of Sutton, photographed with some of his stock in 1959. *The Sutton and Cheam Advertiser*

Today Lens is a legend in the transport book world. John Smith continues to provide a unique service to transport enthusiasts - and is still wearing his beret trademark. *Allan Mott*

Roller coasting

This delightful period piece shows a steam road-roller and wagon on their way to their next job, rumbling along a country road and delaying a Standard Nine on a warm summer's day. The question is, when was the photograph taken? There are several clues in the picture to help you home in on a likely year. See how you get on! (Answer below) ●

Photo: Bill Groome

The fact that the car is driverless should have alerted you to the fact that all is not what it first appears to be. A closer inspection reveals a 'GB' sticker on the rear of the caravan; these national identity stickers were not adopted in Britain until 1954, although elsewhere in Europe they had been used since 1926. Note also the modern 'Long Vehicle' sign. The triangular road-sign on the right-hand verge is of a design outlined in the 1964 Regulations, so we're getting more and more recent. In fact, the photograph was taken in 1992, a delightful pocket of past images moving through relatively unspoilt countryside. Sorry!

Punch and Judy

Freddie Sadler describes a British institution that seems impervious to change - or is it?

Punch and Judy is as old as the hills and as modern as tomorrow. It has the magic, shared with its blood brother pantomime, to cross barriers and we are all children again.

Let me illustrate how Punch is a great leveller. In Covent Garden last summer I was happily watching my friend Percy Press weaving his magic on a very vocal mixed audience. An austere bowler-hatted, pinstripe-suited gentleman fetched up by the booth. He leaned, po-faced, on his furled umbrella and as I watched he unwittingly relaxed, a small smile wiped away his supercilious expression, and like a bursting dam he suddenly shouted out with the kids 'Oompah, oompah, stuff it up your jumper!'

The basic Punch and Judy story has remained unchanged for centuries. Toby the dog, the crocodile, the policeman, the sausages, will strike a familiar chord with anyone who has ever sat on a draughty English beach watching the characters brought to life by a 'professor' behind the scenes. Punch's characteristic voice is produced by using a 'swazzle', a gadget made of an amalgam of bamboo and reed held under the tongue which gives the well-known grating edge to the voice.

Mr Punch has various names. There is Petruska in Russia, Guignol from France, Jan Klassen in Holland, and Pulcinella in Italy. Neapolitan Punch and Judy was first performed in England in Covent Garden by Italian puppeteers in 1662 near Inigo Jones's St Paul's. Pepys recorded seeing a performance. Marionettes were substituted for actors on economy grounds and in the 18th century glove puppets were introduced. Thus it became a two-man show - the owner/presenter and his 'bottler' (collector).

A quick trip down Memory Lane for readers, three generations worth, to say, 'Ah yes, I remember him'. The craft was frequently handed down. So you had Londoners, the Smith family, on their Margate pitch for half a century, until Charlie Smith finished in 1963; Edmonds at Weymouth; Staddon at Weston-super-Mare; Green at Blackpool and Rhyl; Codman at Liverpool and Llandudno; and Maggs at Bournemouth.

> **The basic Punch and Judy story has remained unchanged for centuries**

Percy Press senior started in Swanage in 1935 and was at Hastings from 1948 until his death 31 years later. Percy Press junior, top of the international tree, is the son of this magical maestro. A superb 'professor', he has performed in 24 countries and has his Punch patois in several languages including Esperanto.

'You must establish a bridge immediately you start. It hinges on involvement, the backwards and forwards business. I show my face in the booth, grin hugely, and say "You don't want to see Punch and Judy, do you, boys and girls?" That triggers the eternal running gag: "Oh yes we do", and my "Oh no you don't!"

'I produce Punch who bows three times, then calls "Judy, Judee!". She asks why he called and he says he wants a kiss. "Shall I kiss him, boys and girls? He'll hit me with his stick if I don't. All right, Punch, just one kiss after I've blown my nose on my pinafore." (Raspberry sound) They dance to Camptown Races. Judy fetches baby, Marmaduke. Punch says "Why is he called Marmaduke, children? Because he fell in the marmalade." Punch thumps the baby and it cries. (Grizzle, grunt, groan).'

Familiar faces in our modern days for holidaymakers are Bill Dane, Aberystwyth; Brian Clarke, Lowestoft; Guy Higgins (technical advisor on TV's *Hi De Hi!*) Weymouth; Tony Green at Clacton on a pitch where an incredible character, Claude North, reigned for so long; Wendy Wharam at Swanage; these are a few that come to mind.

There are many others, most members of the thriving Punch & Judy Fellowship, who travel the UK and abroad, appearing at festivals, functions, parks, stores, schools. Dan Bishop, Di Seany, Madam Rose, Stella Richards, Bob Malcome are always in demand. Bob Sacco, a retired regular soldier, is unmistakeable in his special costume, as is Des Turner of Stevenage, in mid-19th-century attire - straight out of Dickens.

One of the yearly highlights is the

Fellowship's Covent Garden September meeting. What a splendid day it is. Visit if you possibly can, with or without children. This eye-catching panorama of Punches is a perfect illustration of the magnetism of the ancient entertainment, even with a few practitioners who have junked tradition and adopted 1993 street cred. There are 30 to 40 booths, each 'professor' with an individual and personal presentation.

John Styles began in the 1950s. 'You can dub me as the reluctant Punchman. As a schoolboy I yearned to be a magician and haunted Oscar Oswald's second-hand business in Baker Street until he took me on as an assistant. While I was earning apprentice fees as a magician, Oscar insisted on teaching me Punch and Judy and initiated me into the mysteries of the swazzle. I practised in Epping Forest and flushed out more courting couples than a Jack Russell unearths rabbits!'

During National Service John's ability to entertain was swiftly recognised and he delighted Service colleagues and civilians alike with his Punch projections. In climbing the ladder John has appeared in a host of TV programmes, enjoying early successes with Marty Feldman, *The Goodies* and *The Avengers*. BBC's *Box of Delights*, starring John, sold worldwide and was beamed on 211 US stations!

North America has also been a happy hunting ground with lecture circuits, campus demonstrations, etc. John has made private appearances for King Hussein and his family, toured the world, and kept the crowds happy outside St Paul's Cathedral when Charles and Diana married. He is an Honorary Fellow of the Royal Society of Arts.

In his inaugural address, John said 'I have learned to respect Punch. This puppet in my hand is an extension of my personality. He is in tatters but when I work with him I give my best performance. I have a very large collection of many ages which includes a few with whom I have no rapport and simply would not use.' John is a very stylish 'professor'.

History was made in the British Embassy in Moscow in 1976. The Ambassador invited 300 guests - remember this is long before Gorbachev - to meet a notorious English scoundrel and

a clique of his unsavoury relatives. Glyn Edwards, a TV director and dedicated Punchman, and Percy Press lined up a dozen 'professors' from around the world. The show was a great success.

The one perceptible change over the past four decades has been the arrival of female operators, brilliantly successful in a masculine domain. Generally they are full of invention, refreshing and, with one or two exceptions, respectful of tradition.

Stella Richards, Essex born and bred, adopted a nursing career and became a theatre sister during the war. 'Punch, dear? The only Punch I ever thought of brought victims to Casualty after the pubs closed.'

Puppetry was literally dumped in her lap. 'A friend came for coffee and brought a ventriloquist's doll. She said "I found the poor thing on a rubbish heap. He looked so mournful I couldn't leave him." She managed to leave him with me, though. I set about getting him into working order and named him Alfie.'

That unwanted gift launched Stella Richards as a top Punch and Judy lecturer, ventriloquist, puppeteer and model-maker, with a collection topping 800. She is proprietor of The Stuff and Nonsense Museum.

The TV series *Hi De Hi!* rang a particular bell with Rose Peasley. 'I'd not been too strong as a teenager and had no set ambitions, so I went to work on the east coast in a holiday camp. Not exactly a Su Pollard, but not far off. I took great lungfuls of air to build me up. I could hardly keep my eyes open!'

There she met her future husband, Eric, who was resident magician. 'Eric could talk the hind leg off a donkey about magic, music hall, circus, Punch - everything. He asked me to be his assistant. What love will do. . .' They married at the season's end. Rose studied balloon sculpture, magic, and using the swazzle specialised in Punch and Judy. Rose became Madam Rosa and deserves her widespread success.

Di Seany was born in Ampney in the Cotswolds and followed her star by going to the London Central School of Speech and Drama. Steady work followed in radio, repertory, and touring. Then came the accolade, a worldwide tour with the Royal Shakespeare

Above **Percy Press Junior with his Dad's last poster. Tommy Cooper smiles overhead. While in the USA on behalf of the British Council, Percy achieved a first when Judy was allowed to break the news to America: 'It's a boy for the Princess of Wales.' He received an instantaneous ovation.** *Freddie Sadler*

Caz Frost (right) is one of the increasing number of female puppeteers, giving the traditional Punch and Judy story a modern feel. She is shown here with Mary Edwards, puppet-maker superb. *Freddie Sadler*

Company. Motherhood meant that she said goodbye to the stage, but founded a puppet repertory theatre with Punch and Judy top of the bill. With husband Jamie, she worked everywhere, even in winter up in the north of Scotland, and taking their show to a handful of children in a tiny schoolroom on the Isle of Skye.

'We kept adding extra scripts such as The Three Bears. My small son gave me the idea of calling my show Judy and Punch. We also appear at quite a lot of clinics to convey medical information to mums and kids through our puppets.'

Caz (Caroline) Frost is a Londoner who wanted to be an actress. She had no doubts, but the uncertainties of the profession worried her family, so she was persuaded to take a teacher training course in drama as a safety net.

'At Exmouth College each student had to make a personal presentation for final assessment. I opted for Punch and Judy. It had begun to fascinate me and I passed the test with high marks. I left college, took an enormous plunge, and soloed with a summer season in the West Country.'

Constant study, experience in gauging audience reactions and advice, freely given by experienced puppeteers, kept Caz on the move, and surviving bad patches 'helped me to believe in myself, which is all important'. She joined The Street Theatre and met and married Mike, 'Major Mustard'. Caz is traditional but has brought a feminine slant to the former male province.

'I am not a mad feminist with tunnel vision and a tendency to chuck the baby out with the bathwater. My Judy is a proud mum, but not the brow-beaten creature usually portrayed. She is proud of her child, and it is a modern kid. Judy goes hammer and tongs for Punch when he bullies her, and is physically overcome in the end after she's given him a black eye. I have made other changes in the characters, nothing outrageous, but giving a truer balance, I feel.'

We are all hooked on, even respond with affection to, the appeal that this irascible character projects. Mr Punch, let's fact it, is a wife-beater, thief, liar, cheat, bully, con man and a murderer. Magic? On one Sunday there is a special service in St Paul's, Covent Garden, when he is up in the pulpit. And why not? He is remembered with plaques honouring actors in that church.

As such he should have the last word. 'That's the way to do it!' ●

Maestro John Styles and friends in Covent Garden. *Freddie Sadler*

The Orangery, Kew Gardens

The fascination of the Royal Botanic Gardens at Kew, including its Orangery, has attracted approximately a million visitors each year this century, with a peak of about 4 million in 1915. During the first half of the century the admission fee was 1d. A three-fold increase in 1951 took it to 3d. After a number of increases during the 1980s it reached £3.30 in 1990, a figure more in keeping with many other fine buildings and parks.

The Orangery at Kew has housed not only oranges, but also wood, books, maps, gateau and sirloin steak at various times during its 200-year existence. It was originally designed and built in 1771 by Sir William Chambers, architectural tutor to King George III. The sole purpose in those days was to over-winter orange and other citrus trees. This was again its purpose in the 1960s, but the building suffered from a recurring problem of the effects of the damp atmosphere on the wall plaster.

For nearly a century, however, from 1862 to 1958, it housed an exhibition of timbers transferred from the Great Exhibition, which was described at the time as 'a magnificent collection of timbers, cabinet and furniture woods'. In the 1950s, however, the Orangery presented a drab scene, as can be seen from the accompanying photograph; it still appeared to be - as indeed it was - an antiquated Victorian museum.

What a contrast today! Although the exterior of the Orangery is practically unchanged from when it was built, peel off the outer skin and taste the difference. The inside is

split into two parts - one is The Orangery Restaurant, and the other slightly smaller section is a gift shop, with all the expected trinkets - postcards, china, pictures, tea-cloths, jigsaw puzzles, scented soap and pot-pourri.

The Orangery has been more recently described as 'surely one of the prettiest buildings to house a restaurant anywhere in Britain'. The garden setting is tranquil and through the tall windows are delightful scenes in every season. Here visitors to Kew Gardens can now enjoy morning coffee, lunch or afternoon tea with orange marmalade. The restaurant is open seven days a week; currently it is self-service, although for a time waitress service was offered.

Forty years ago the Orangery was a mausoleum - today it is alive and inviting. ●

Modern exterior and 1950s interior photos courtesy of Royal Botanic Gardens, Kew; 1992 interior and notes by Dave Radford

Past and Present

Home improvements

Dora Tack lived at Grange Farm Cottage in the 1950s - 40 years on it presents a different face to the world.

Grange Farm Cottage in Hilton, Huntingdonshire, was a tied house belonging to farmer Joseph Leycester, for whom my husband Frank worked. In season, Frank drained and dug ditches, laid live hedges, chopped and singled sugar-beet, all at piecework rates, in between the normal farm work and involvement in the long days of 'Harvest Month' each year. Over the years the house has undergone various renovations, but remains a delightful family home to this day.

The first photograph shows the cottage as it was in 1952. There was a beautifully carved beam extending from front to back in the large room to the right of the front door. On the left of the cottage the rain-water tank was fitted with a wooden cover, held in place by

bricks after a cat fell through the previous sacking cover and drowned. There was no letter-box cut through the thick front door - a box had to be fitted on the wall for mail and newspapers. When the cottage was re-roofed, a small cubby-hole was found behind the wall to the right of the chimney in the large bed-

room (top left in the photograph); access to this concealed space was through a hole in the brick wall of the cottage, well hidden near the roof of the adjoining cart hovel. A piece of tattered carpet and a torn blanket indicated that someone had at some time hidden or slept there.

Forty years on the house is now called simply Grange Cottage and is privately owned. A new extension has been added to the left of the cottage, extending past the old bathroom window, with a large living-room window and a side door, and windows where the old cart hovel used to be. Maybe some of the tiles on the roof were once part of the sheds and stables which used to surround the strawed yard at the back of the cottage when it was Grange Farm Cottage and had bullocks in the yard during the winter. The new colour-wash covers up the iron reinforcements at the corner of the cottage on the right. The old front door is still in use, but now has a much-needed letter-box and knocker; the picket fence has gone and has been replaced by a hedge; and the water tank has gone too. ●

1952 photograph by Dora Tack; 1992 photo by Allan Mott

The road to Silloth

These photos show that even in the most remote areas changes still occur. This road junction is in the far north-west of Cumbria, between the towns of Silloth and Maryport. As can be seen, the route has been upgraded over the years from a single-lane to a two-lane road, bordering the Solway Firth. Surprisingly, however, the old-style signpost has been retained, although it has been moved to the seaward side of the road from its original position. The road marker posts, so prominent in the old photo, have also been retained, but the peeling paint means that they are now starting to show their age. To accommodate modern traffic levels (although this area's remoteness means that such levels are still low), the junction has been completely rebuilt and tarmacked. But despite all these changes at this isolated spot on the Solway Firth

with its miles of deserted beaches, the distant hills of Scotland still dominate the scene as they have always done. ●

1972 photo from the Peter Robinson Collection; 1992 view and notes by John R. Broughton

'We said we wouldn't look back. . .'

. . .but Polly Bright does, to recall her student days at Bristol University

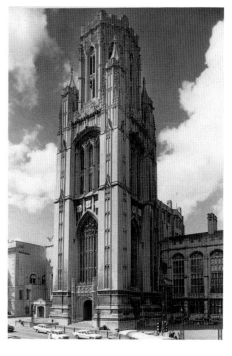

The imposing tower of the main building of Bristol University. *Bristol University Information Department*

*B*ack in the early 1950s, when I went up to Bristol to read English, sugar, butter and various other foodstuffs were still on ration, supermarkets and canned drinks hadn't been heard of, and, unless you counted ice-cream, neither had frozen food. There were no computers, credit cards, motorways, launderettes or jeans. The only broadcasting authority was the BBC with three radio services and just one black and white television channel operating mainly in the evenings.

Not that this affected us much as few students possessed or felt the need of a portable wireless, and we were only too aware that academic mentors considered television rather *infra dig*.

We were stylish if unsophisticated, and when I alighted at Temple Meads station it was only the second time in my life that I had travelled alone by rail. I carried a handbag and weekend case, my one large suitcase having gone ahead luggage-in-advance. Its contents were fairly typical - a hot-water bottle, *The Complete Works of Shakespeare* and my clothes.

What most of us also arrived with was a feeling of exhilaration that by some amazing stroke of luck we'd found our way to one of the best parties in the world, which, provided we kept our noses clean, would last three glorious years.

If competition for places was high -

and with only seven per cent of the population going to university it must have been - we were not unduly stressed by it and once there the name of the game was *enjoy yourself*. True, medics, vets, dentists, engineers and scientists signed in for lab work and lectures and generally had their noses kept to the grindstone, but in the Arts Faculty the attitude was 'We don't insist you attend lectures, but we do insist you pass examinations.' This we contrived to do without letting it interfere too much with other pursuits.

> *. . . first-year women students were not allowed out after dinner . . .*

As the age of majority was then 21, both wardens and landlords were *in loco parentis* and the halls of residence were single sex. At Clifton Hill House, first-year women students were not allowed out after dinner without special permission, and even then they had to sign themselves in and out like parcels, state where they were going and be back by 10 pm. Second and third years didn't need to seek permission, but were otherwise bound by the same rules. These were relaxed for Saturday dances, when a

senior student held a key and counted everyone in. On Wednesday and Sunday afternoons they were allowed to entertain male visitors in their rooms, which was devilish cunning as Wednesday was always devoted to sport. None the less it was considered liberal, as rumour had it that at another university a man couldn't enter a girl's room until the bed had been moved outside into a corridor.

None of this was resented as most had to toe the line at home and similar rules applied in other hostels like the YWCA. Terms like 'sexism' and 'double standards' hadn't entered the language and nobody begrudged men the freedom they enjoyed in their halls. At Wills there were still a handful of post-demobilisation students working their way through the system, and no warden was prepared to impose petty regulations on a former pilot or submariner who had done the state some service. Of the younger intake, some already had two years' National Service behind them, and generally men were judged quite capable of taking care of themselves.

When Prince Philip visited the university during the 1950s, students in the Clifton Hill House Library were wearing academic dress as a matter of course. Today gowns are only worn on formal occasions. *Both Bristol University Information Department*

Then, as now, accommodation was at a premium and first-year students were not automatically offered a place in hall. Having started in digs, I stayed outside, finding hall fees beyond my means anyway. The present mixed university flats and houses with self-catering facilities were beyond our wildest dreams. Landlords registering with the Accommodation Officer then provided board as well as lodging and could take girls or men, but not both. Those with unmarried daughters or motherly types wanting someone to spoil usually opted for men, while those plumping for girls often did so because they thought they would be less work.

In my first digs I met Jennifer, a science fresher, who became one of my closest friends. We found ourselves in the bosom of a po-faced family who were as hot on curfew as they were cold on comfort, but claimed that they only took students to help the university. Everything was ersatz, from our landlord's smile to his wife's cooking, but my not taking sugar at least put an end to their practice of putting saccharine in the teapot. On Sunday morning, after preparing a rice pudding with equal quantities of milk and water, our landlady gave us an egg, the only one I remember, and bade us share it for breakfast, leaving us to work out how to cook and divide it.

On the accommodation list her offer to put students' laundry through her newfangled washing-machine for 'just the price of the soap-powder' had looked generous. I kicked off by taking her a packet of Tide, a new wonder detergent often in short supply, but I still got charged for laundry, and we didn't need to be Einstein to deduce that our laundry paid the cost of the washing-machine. Early in our second term they got shot of us, reporting us to the Accommodation Officer for sneaking upstairs in the dark, shoes in hand, a few minutes after curfew.

Terrified of being sent down we took refuge with Auntie, a septuagenarian spinster who kept house for her widowed brother-in-law. Not for the world would Auntie have split on us, but old habits die hard and as we crept past the hall cuckoo-clock in the dark on Saturday nights we were never sure if a single 'cuckoo' signified 11.30, 12.30 or the outrageous hour of 1 am. Although she was kind and welcomed boyfriends, her terms were expensive as they included lunch. We moved on with regret because it was too far and too much hassle to go home at midday and we couldn't afford to skip lunch and buy another in the refectory.

We were in rapport with Sadie, our third landlady, from day one, and remained friends until she died. Even though the bathroom geyser backfired horrendously on ignition and I had to move my bed to dodge snow falling through a sash window that wouldn't close, we still thought we were on velvet for we were nearer the university and Sadie was fairly young and easy-going.

Although we had no Preview Days, the Freshers' Conference was established and in between talks, guided tours round departments, welcoming dances, auditions for freshers' plays and just making new friends we trotted off to Ede & Ravenscroft to get kitted out with university scarves and academic dress.

At that time, without 'a black, stuff gown of the approved pattern' arts students couldn't so much as set foot in the library or a lecture or tutorial room, all of which were housed in the main building, now the Faculty of Law. In the morning we pulled them out of locker cages and, weather permitting, flapped about in them all day like grounded blackbirds - round the shops, over to Forte's Ice-cream Parlour, where we sat in basket chairs at glass-topped tables drinking coffee and catching up on the latest gossip, and up and down to the Victoria Rooms, which was then the Students' Union and always affectionately referred to as the Vic Rooms.

Those in hall had to remember to take their gowns home for formal dinners. These took place twice a week at Clifton Hill House, when grace was said and students took it in turns to sit with the Warden at High Table. Other halls had other customs; Wills simplified matters by insisting on gowns at every meal. Corporate feeling was strong and our gowns were both our pride and our comfort. We snuggled inside them for warmth in draughty halls and wiped our pens on their generous sleeves.

For ceremonial occasions we wore the full works - gowns, black suits, stockings and shoes, and white blouses. While men got away with any dark suit and enjoyed the dignity of mortarboard and silk tie, girls were only permitted choristers' caps and ties in dreary machine-knot.

Our first ceremony in full regalia was Matriculation, which acknowledged that we were qualified to enter university, even though we had been clattering round the place for several weeks. Those reading English had to have Latin as well as a foreign language, which probably helped whittle down the numbers applying.

Soon I was donning my finery again for another ceremony in which honorary degrees were conferred by our Chancellor, Sir Winston Churchill. He'd held the office since before many of us were born, and when he was in power he never let his duties as Prime Minister get in the way of it. He himself collected honorary degrees, honorary fellowships and

freedoms of cities like other people collected stamps, and could deliver a speech the like of which you would never hear again. The only way a student could get into the ceremony was as an undergraduate usher and when I won a place in the ballot I was treading on air. Carrying a staff, I filed in procession into the Great Hall, sat through the speech too excited to digest it and formed up again afterwards in a guard of honour. It was a high spot and meant far more to me than my own degree ceremony two and a half years later.

By today's standards we were dressy, turning up for lectures in high heels and nylons with swishing skirts and hand-knit jumpers under our gowns. To a man our opposite numbers wore shorts and ties with blazers or sports jackets and the ubiquitous flannels. Lecturers went in for three-piece suits under silk gowns, so at a glance you could tell you was who. We called them Mr, Miss, Doctor, Professor or Sir, while with equal courtesy they always address us as Mr or Miss. They were unfailingly kind, and far from creating barriers the reciprocal courtesy brought us close together.

> ... we were dressy, turning up for lectures in high heels and nylons with swishing skirts ...

Every generation produces its star performers. One such was Mr Gifford, timetabled to deliver a double faculty lecture in the Reception Room form nine until 11 on Wednesday mornings. To hear him on Dickens was better than the pictures, but after $59\frac{1}{2}$ minutes he literally had to take a breather and would bounce down from the dais and fling open a casement window just as Great George began to strike 10. Then, while he leant out to take great lungsful of fresh air, we clutched our gowns to us in the howling draught and tried to field our flying notes.

Basil Cottle, later of *Dictionary of*

English Surnames fame, thought nothing of dressing up in drag to join us in the English Department Christmas Show. Glynne Wickham, after whom the present studio theatre is named, broke the sartorial mould when he politely suggested that ladies might find it convenient to wear slacks for his next Drama practical. That created a stir as those of us who possessed a pair hadn't thought to pack them unless we were involved in sport. Getting them dispatched from home presented no problem. In those days of one-tier post a letter sent early in the morning would arrive by the afternoon mail and it was possible to get a reply the next day. Drama was a popular choice for Single Honours students, who took two subsidiary subjects in the first year in case they fluffed examinations and had to transfer to a general arts course. Once in a while we worked with fledglings from local theatres. The Bristol Old Vic was the birthplace of the '50s hit musical *Salad Days* about two young graduates who poignantly sang:

'I'll remind you to remind me
We said we wouldn't look back.'

To us the name Ted Heath was not associated with a politician but a dance-band leader, who often played for Saturday dances and balls at the Vic Rooms. For dances we wore cocktail dresses, which were then cheap and cheerful, while our ballgowns were often run up at home or by local dressmakers. Many of the men's dinner jackets were hand-me-downs and signs of age like moth-holes added a certain cachet. Jive had gone out of favour and it was all ballroom dancing, apart from an odd medley of Hokey-cokey, Palais Glide and Bumps-a-daisy thrown in for fun. A tiny minority used the bar, the rest being content with coffee in the interval. We also danced to records at the Fry-Haldane Tuesday Hop for students living at home or in digs, and at the International Society on Sunday afternoons.

There was no shortage of live entertainment with plays put on by the Fry-Haldane Society and the various halls,

Rag Week was and still is the high spot of activity and fund-raising for university students. The lorry in the 'past' view, with its distinctly war-surplus look, carries Wills Hall's 'sound barrier' theme; first broken in 1947, the sound barrier was probably still hot news when this photograph was taken.

The mode of transport in the 1990s is different, but the aim is still the same!
Bristol University Information Department/Bristol United Press Ltd

a revue called Radio Wills, entertainments dreamed up by the departments, and Dramatic Society and Drama Department productions. Activity peaked in Rag Week and later in Union Week at the end of the academic year, a time of post-examination festivity that has not survived.

It all happened on a shoestring. Only four students owned cars and just a handful more had motor bikes. Most of us worked in our local post office at Christmas and at any menial job we could find during the long vacations to keep ourselves in books and clothes.

For all the high jinks, I remember only two cases of couples 'having to get married'. If the Students' Health Service had offered contraception counselling and provision in the '50s, parents would never have let their daughters leave home.

Students have always had to work to get results, but what today's might envy was our lack of concern about the future. There was full employment and although salaries were modest and promotion slow, those on vocational courses had no difficulty finding work in their chosen professions. It was always trickier for Arts graduates, particularly girls, but everyone found some sort of niche, and when we look back the verdict is always, 'Didn't we have fun!' ●

No 2: The White Swan, London SW1

Anthony Phillips

1965 Fleet Street had its 'slopers' by the dozen - The Cheshire Cheese, the Printer's Devil, The Punch Tavern included - where newsman would 'slope off' for well-earned refreshment. The White Swan in Vauxhall Bridge Road, Pimlico, was the 'sloper' for council people working at Middlesex House (of which the Swan formed a part), medics from the Military Hospital on Millbank, Tate Gallery staff and visitors, businessmen, and the odd nurse or two from Westminster Hospital who probably couldn't afford the price of a Babycham, let alone the pints of Watneys Special and Red Barrel which were in then a 2s 6d a pint.

Built just before the Second World War, the White Swan had four bars with an overall plum colour effect. No carpet in the saloon, but a chessboard lino floor. To brighten up the Pimlico setting, however, there were swaying palm trees on the walls. But the lounge was softly lit and spacious, and there was a real grand piano which was strangely devoid of beer-glass rings.

Is it that long ago that we were belting out '24 hours From Tulsa', doing a job on 'Please Please Me', and thrown out for merely asking for a *cuba libre*? ('Coming in here with your bleedin' Costa Brava ideas. . .')

The veal and ham pie with salad wasn't bad for three and a tanner and there were pioneer attempts at a vindaloo. There were heart-stopping moments at the bar billiards table at times as one of those mushroom-type 'thingies' quivered before pirouetting down the hole. ('Pool, mate? Great Smith Street baths up the road!')

1993 O my word, haven't we gone all pound-noteish in the Vauxhall Bridge Road? A real eighteenth-century boozer with the latest karaoke, too. Very continental those little tables on the pavement. The lady from Texas was not amused when someone told her that she was sitting at the very same spot where Ted Rutter, the concrete inspector, shot the cat - eight pints of black and tan, thirty year ago. The C registration Wolseley parked outside in 1965 has been replaced by a J reg Sierra.

But to give credit where credit is due, they haven't done a bad job on the conversion. They've torn the wallpaper and plasterboard off the walls, and ripped out the ceiling board to reveal the original 5 x 12s. The scrubbed table look is in, and the young staff at least have a clean white shirt every day. The food's quite trendy, and to get a piping hot cup of coffee in a London pub is not a bad idea. The Tate Galleryites are there with their Monet brochures, along with publishers from Random House (formerly Middlesex House), and there is the *thumph, thumph, thumph* of Madonna, Michael Jackson and Queen in the evening. Now I ask you: how do you sing along to 'Bohemian Rhapsody'? *Mama mia!* As regards singing, Karen Carpenter knocked them all into a cocked hat. ●

1965 photograph by David Keane; 1993 photograph by Anthony Phillips

---

URBAN DEVELOPMENTS

A recent report by the Council for the Protection of Rural England has highlighted the enormous spread of towns, cities and roads since 1945. Almost 2 million acres of countryside has been lost, 1 million acres being farmland. That's an area equivalent to Berkshire, Hertfordshire, Oxfordshire and Greater London combined, lost under urban spread and motorways. According to the CPRE, the current rate of 'countryside loss' is about 25,000 acres a year, and by the middle of the next century about 20% of England will be urbanised - the present figure is 15%.

How has your area fared?

Yorkshire and Humberside is the worst hit area, showing an 88% increase of urbanisation since 1945.

East Midlands is next, with an increase of almost 75%. Derby, Leicester, Northampton and Nottingham are growing most, and 190,000 acres of farmland has been lost.

Close behind is the **South West** at 70%, where well over 300,000 acres of moorland and rough grazing (including parts of Dartmoor and Exmoor) have gone.

West Midlands is up 68% - 200,000 acres of rural land has been lost, an area twice the size of Birmingham.

The **North West** (66%) is now the most built-up area of England, and has lost more farmland relative to its size than anywhere else.

In the **North** the urban area has increased by 54%, equivalent to the area of Tyne & Wear.

Urban area in the **South East** is up by almost 470,000 acres (44%), nearly twice as much as any other area, and twice as much farmland has been lost than in any other.

Finally, in **East Anglia** farmland covering eight times the area of Norwich has been lost since 1945.

Overall, the trends are 'extremely worrying', said the CPRE. Have any of you been recording it with photographs?

Oil change

David Kinchin on the packaging of 'liquid engineering'

Castrol may well lead the field with their own brand of liquid engineering, but the modern motorist is often bamboozled by the term 'viscosity'. Do we need 15W/50 oil or perhaps a fully synthetic 10W/60 oil which is sold under the title 'Formula RS'. . .

Gone are the simple days of allowing the petrol station forecourt attendant to dive under the bonnet, check the oil, smile knowingly and return from a trip to the workshop with a metal jug containing a pint of his very best motor oil. Gone are the forecourt attendants, gone is the pint, and in their place we have self-service and things called 5-litre cans.

The packaging has changed considerably too. Forty years ago, oil and petrol companies sold their products in strong metal cans, a gallon at a time. Names like National Benzole, Shell-mex, and Royal Daylight Oil were proudly embossed on the sides. These cans have now become very collectable, and when restored to their original colours are very pleasing to the eye, and the environment.

The modern-day replacements are less colourful, very oddly shaped, and made of plastic. What is in the can, however, is vastly improved. It is purer, a more flexible 'engineered' product which lasts for many more miles than the oils of the 1950s and '60s. The price has also naturally increased considerably over the years. What costs between £8 and £10 today would only have cost around 4s 6d in 1952. ●

Photograph of old petrol cans by David Kinchin; 1993 photograph of oil 'cans' by Julia Thorley

Obtaining the correct Grade

In ordering your oil be careful to state the make as well as the grade. For example, never ask for XL, A, ' Double ' or ' 30,' but always ask for **Castrol XL, Mobiloil A, Double Shell, Essolube ' 30 ' or Price's Motorine ' M,'** according to the brand chosen and see that the oil is drawn from a container bearing the well known trade mark.

From the Standard Eight Instruction Book, 1945-8

Action! From cine to camcorder, Allan Mott reviews the development of

Home movie magic

From an early age during the late 1940s I realised that my father was a keen cine enthusiast. It was only later that I realised he must have been one of the few.

My father's first projector was a Pathescope model called 'Baby Pathe'. Introduced before the Second World War, the original machine was hand-driven and was for 9.5 mm films. My father saved up and bought one, to which he later added an electric motor which made things easier and smoother on the screen. 9.5 mm was a very good quality film; its central sprocket hole between frames enabled more space for picture area and quality. Sadly, this format was not to last, being replaced by 8 mm and 'Super 8' size films. Since the appearance of machinery, history is full of instances where the odd-man-out has been swamped by what, rightly or wrongly, has become the accepted standard. In recent years the battle between VHS and Betamax video systems was won by VHS, even though Betamax provided the better quality. So it was with cine film - 8 mm film

Above right Dad at work on a holiday feature at Westbrook in 1950, using his 9.5 mm camera. *Allan Mott*

Right In 1969 an amateur film-maker stands poised and ready for action with an 8 mm zoom camera, the '60s equivalent of the camcorder. *Allan Mott*

Below right Still convenient and easy to handle, but far more sophisticated - a modern JVC camcorder. *Julia Thorley, with acknowledgements to Dixons Ltd*

Below The 'Baby Pathe' cine kit - projector, film, camera and screen. *Allan Mott*

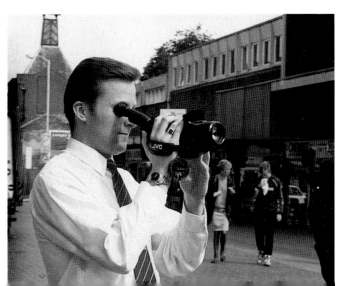

was easier to produce and conformed to the standard.

In addition to processing your own films as my father did, with chemicals, darkroom and running water, films could be hired from a small number of libraries, and films that had been on hire for some time occasionally came up for sale or auction. I remember my father had the first reel of the film classic *The Cabinet of Dr Caligari* which he bought at an auction, other bidders getting the other reels. Swap meets were not around then, but it would have been an interesting scene: 'Has anybody got the last reel of. . .?' '*Birth of a Nation* reel 4, anybody. . .?'

One of the problems with film as opposed to video is time. Whereas a video runs for two or three hours, or even longer, a film is usually 20 minutes per reel, and in the case of a long feature this was frustrating, not only because of the disruption caused by turning on the lights while the reel was changed, but also because viewers tended to lose the thread of the story, let alone the emotion involved in following the film. To get round this, features tended to be abridged versions of the original, or one-reel comedies or cartoons. Incidentally, one of my father's one-reel films was *Stage Struck Donald*, in black and white and the first film in which Donald Duck starred.

As most of the home movies were silent, sound being expensive, prohibitive and unavailable to the amateur, the spoken parts were captions in blocks, not subtitled as today. Necessity being the mother of invention, in 9.5 mm (but not the other formats) it was possible to notch the film, and this held the caption on screen while it was read. This method also saved film; if it was all titles, at 24 frames per second there would not be much room for pictures in the reel. In fact, most home movie projectors were set for only 18 frames per second.

During the 1950s Dad converted our loft into a purpose-built cinema, to which I added my model railway - he wasn't going to have all the room! Cinema and railways can and do go together well.

With the new cinema came new equipment, an 8 mm and 16 mm silent Speco projector. To the latter my father added and modded a sound system. Being experimental it was prone to teething problems, especially feedback, which produced very loud 'wows' and hums. What the neighbours thought has not been recorded, but my moth-

Top This advert from *Amateur Cine World*, November 1956, shows just three of the cinematic delights promised for the forthcoming year.

Above Today a vast range of feature film videos can be hired from every corner shop, as well as the specialists. *Allan Mott*

er's frustrations were well heard, especially at dinner time.

'Charles, your dinner is ready!'

Wow. . . Hum. . . Another cold dinner for Dad.

Later a Bell & Howell L516 projector was added. This was built for the RAF and other government services, and, being very robust, was used in all climates and environments for showing educational as well as entertainment films. When the model was being replaced by a more modern machine, a large number were sold under Government Surplus. Films could still be hired but from different libraries. The larger film distributors of J. Arthur

Rank and Columbia, amongst others, had some very interesting catalogues. There were also free libraries showing promotional films, such as those from BP, Sound Services and British Transport Films. I see that today Channel 4 still shows some of these very interesting films.

Sadly in 1981 lightning struck our house, and in the ensuing fire not only the equipment was lost, but also many, many happy memories: Croydon Airport. . . Crystal Palace the day after the fire. . . my wedding film. . .

Modern home movies are now recorded on videotape - not so many cuts, if any, and full soundtrack. One of the most noticeable differences between film and video is, of course, the size of the screen, film providing the larger picture. An interesting observation is that TV operates at 25 frames per second, but the sound is still 24 frames - what has happened to the other frame!

The modern camcorder offers features undreamed of in the age of cine - instant play-back through your TV (or in some cases through the camera itself), automatic exposure and sophisticated zoom facilities, and stereo sound. There's even a video 'walkman' using small 8 mm cassettes. Equipment is also available that can transfer your old 'Super 8' memories on to video, so the past does not need to be lost or forgotten. ●

FLASHBACK - THE TOP THREE UK BOX OFFICE HITS OF THE LAST FOUR DECADES

1950
1 *The Blue Lamp* (Dirk Bogarde, Jack Warner)
2 *The Happiest Days of Your Life* (Alastair Sim, Margaret Rutherford)
3 *Annie Get Your Gun* (Betty Hutton, Howard Keel)

1960
1 *Doctor in Love* (Michael Craig, Leslie Phillips, James Robertson Justice)
2 *Sink the Bismarck!* (Kenneth More, Dana Wynter)
3 *Carry On Constable* (Carry On team)

1970
1 *Battle of Britain* (Laurence Olivier, Robert Shaw, Michael Caine, Kenneth More, etc, etc)
2 *On Her Majesty's Secret Service* (George Lazenby, Diana Rigg)
3 *Butch Cassidy and the Sundance Kid* (Paul Newman, Robert Redford)

1980
1 *The Empire Strikes Back* (Mark Hamill, Harrison Ford)
2 *Kramer vs Kramer* (Meryl Streep, Dustin Hoffman)
3 *Star Trek - The Motion Picture* (William Shatner, Leonard Nimoy)

1990
1 *Ghost* (Patrick Swayze, Demi Moore)
2 *Pretty Woman* (Richard Gere, Julia Roberts)
3 *Look Who's Talking* (John Travolta, Kirstey Alley)

Prices table

When you look back a few decades at the prices of things, it's easy to think that just because everything cost less then, it was cheaper than it is today - but *in real terms*, of course, many things are in fact *cheaper* now, relative to our disposable income.

To help *Past and Present* readers get an idea of relative prices, we have put together this table, showing how wages and the prices of several staple products have changed over the decades. This is of necessity only a rough, 'round figures' guide, but it is interesting, for example, to see that while a Mars bar has gone up 10 times and a pint of bitter 20 times, wages have gone up some 36 times!

The earlier figures have been converted to the nearest decimal equivalent to make the comparison easier.

	1950	1960	1970	1980	1990
Average weekly wage	£7.29	£14.10	£25.90	£109.50	£263.10
Pint of milk	2p	3p	5p	$16^{1}/_{2}$p	32p
Pint of bitter	5p	6p	24p	$35^{1}/_{2}$p	£1
Average white loaf	$2^{1}/_{2}$p	5p	11p	34p	53p
Mars bar	2p	$2^{1}/_{2}$p	3p	14p	21p
Daily Telegraph	0.6p	1p	$2^{1}/_{2}$	12p	35p
B/w TV licence	£3	£4	£6	£12	£24
Road fund licence (private car)	£10	£15	£40	£60	£100

£sd/decimal conversion
$2^{1}/_{2}$d = 1p
6d = $2^{1}/_{2}$p
1s = 5p
10s = 50p
20s = £1

Open wide!

Former dentist Bernard Leton peers into the nation's mouths and sees vast improvements

Mention of the dentist still causes fear and cringing among some people, but modern technology is gradually taking the dread out of that six-monthly check-up. Perhaps the day is not far off when a visit to the dentist will become an actual pleasure - indeed, something to look forward to.

During the war and for some years afterwards sweets and chocolates were rationed; they were in short supply and could only be bought on coupons. Although this might have been hard to live with, it had the beneficial effect of reducing tooth decay. (Sweets produce the acid which attacks tooth enamel, allowing bacterial invasion and eventually dental decay.) During those years teeth suffered less from dental decay and fewer extractions and fillings were necessary.

In my student days, dental decay in children's teeth was rampant. Most children lost many teeth before the age of six and this caused malformation of permanent teeth in later years. Today it is rare to see dental decay in baby teeth and mass extraction is a thing of the past. Fifty years ago most people expected to lose all their teeth by the age of 40. Today there are many 80-year-olds about with most of their natural teeth still intact.

Hardly a child in the United States goes about without braces on his or her teeth, to straighten them and cause them to have a desirable and regular alignment. The idea is now catching on here; more and more children are having orthodontic treatment, as it is called, and braces in youngsters are becoming commonplace so that future

Above left This picture of an old-fashioned foot-operated electric drill will no doubt bring back many a painful memory for some of our readers! *Bernard Leton*

Left Perhaps the most remarkable of all dental apparatus inventions is the high-speed drill. This revolves under air pressure at up to 80,000 revs per minute, and a tiny diamond drill cuts the enamel like butter. It causes hardly any discomfort, and to keep the temperature down a jet of water is sprayed on to the working area during its use. Incidentally, the gloves and mask worn by the dentist here are not a legal requirement, but have become a sensible precaution. *Julia Thorley, courtesy of John C. Gray*

generations can have perfect arches.

As in other areas of health care, there is now much emphasis on prevention rather than cure. Regular visits to the dentist can help stop problems before they start, and there is now a vast array of over-the-counter products that can be bought to ensure a healthy mouth. Years ago you could buy toothpaste and a brush and that was it. Now you can buy toothbrushes in all shapes and sizes, a variety of toothpastes, plaque-disclosing tables, dental floss, and antiseptic mouth-washes. Fluoride has been another great step forward in the control of dental disease; it is now incorporated in toothpaste, added to drinking water in some areas, and can be taken in tablet form. Education of the public has taken great strides forward, and proper dental hygiene is now accepted as part of maintaining general good health.

The fear of the dentist is gradually declining because of the use of new anaesthetics. Local anaesthesia can be used to combat the pain of the drilling of teeth, but some people are still scared of the needle. Other methods can be used, such as nitrous oxide and oxygen given through a mask. Some dentists will even offer to hypnotise particularly nervous patients.

Fillings have for many years been constructed from silver amalgam, and there is still no alternative that is 100 per cent acceptable. It is safe, durable and satisfactory in every way, and continues to be used worldwide. New white materials are being used and their appearance is better, but they lack the permanence of amalgam.

Should all else fail and you end up having to wear false teeth, even this is not the ordeal it used to be. Until 1945, because of the war there had not been many advances in dentistry but there was one noticeable change in the materials used for the construction of dentures. Rubber (vulcanite) was unobtainable during the war, so we moved to making dentures in acrylic or plastic - we have never look back. Acrylic is vastly superior to vulcanite in appearance.

Crowns and bridgework have taken enormous steps forward. Since the introduction of the high-speed drill, teeth can be prepared for crown and bridgework in a very short space of time. Crowns are more lifelike in appearance and are in acrylic instead of the old porcelain which was brittle, glasslike and unnatural in appearance. The precious metals gold and platinum are used as bases for bridgework, and it is remarkable how missing teeth can be replaced with permanent bridges to look like a mouthful of natural teeth, with the advantage of a permanent fix - no more removing of dentures after meals, no more slipping plates causing embarrassment, and no more raspberry pips getting under them! And bridgework can be cleansed by high-speed jets of water attached to taps and moved around the inside of the mouth. Adhesives are vastly improved - crowns fixed in are very per-

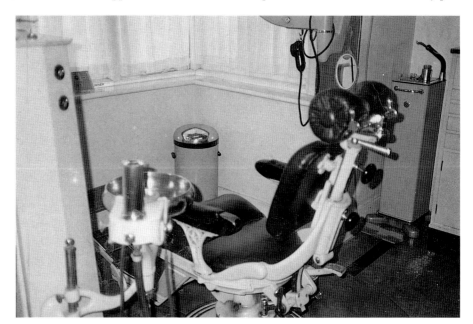

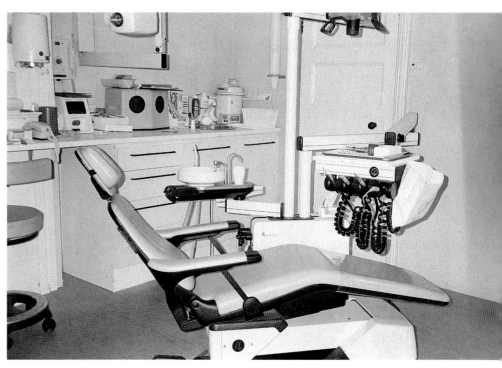

Above right This chair provided sterling service in its day, but is a museum piece by today's standards. Note the old spittoon and waste receiver in the background. *Bernard Leton*

Right Patients do not sit upright any longer - today's chairs are electrically operated and can be laid as flat as an operating table. The dentist sits at his work behind the head of the patient and all the apparatus is at his fingertips. Apart from the chair, much of the equipment used in 1993 is the same as that from 40 or 50 years ago. One of the most notable improvements, however, is the introduction of the steriliser unit (in the centre of the picture). During the 1950s many dentists used only boiling water for cleaning their instruments - or simply did not bother! *Julia Thorley, courtesy of John C. Gray*

manent, and bridges do not come adrift.

Implants are another step forward in dental prosthetics. Plastic teeth are actually implanted into the bone of jaws where they take root, as it were, and replace unsightly gaps. There is no rejection such as occurs in other parts of the body when tissues are implanted.

One of the greatest problems in dentistry has always been the full lower denture. There is no suction, the tongue gets under it and dislodges it, and the lips tend to push it adrift. However, vast strides have been made to solve the problem and the method now is the implantation along the ridge of the lower jaw of a special metal strip. This is fixed into the bone and covered with the soft mucous membrane, then left until the whole area heals. Attached to this metal strip are two upright posts in the region of the lower canines. These stand rigidly above the gumline and on to these posts a full lower plastic denture can actually be clamped or slotted. There is no more mobility, no more coming adrift. Results have been spectacular.

There have also been improvements in root canal treatments. Even if all there is left in a lower jaw are the two canine roots - the strongest - these can be preserved with root canal treatment and used as buttresses for fixing the full lower denture.

Hare lip and cleft palate, unsightly disfigurements 50 years ago, are things of the past. Cleft palate can be dealt with expertly by modern oral surgery and hare lips are so well treated that it is impossible to spot the scars on the upper lip.

X-rays have improved enormously in the last five decades. Now the whole head can be x-rayed in one operation and all the teeth are shown in one picture. Spotting the areas of decay is more easily pinpointed and areas of regression can be seen too. Elsewhere, investigations are being made into the use of laser beams in dental surgery. There are new drugs to combat gum disease, there are new mouthwashes and dental pastes. The search and the improvements go on.

And all this improvement in dental health is in spite of the fact that we are the second biggest sweet-eating nation in the world. Only the Swiss eat more sweets than we do! ●

Forty years of breakfasts: Kellogg's milestones

Ever since the 1920s, when Kellogg's products were first sold in Britain, they have been a prominent feature of the nation's breakfast table, and the benefits of a wholesome, nutritious start to the day have always been emphasised. In 1938 the first British plant opened.

1947 Corn Flakes back in production after the war

1949 First free gifts - picture cards

1951 Rice Krispies (introduced in the US in 1928) back on the British market

1952 Bran Flakes introduced; 100,000,000 lbs of Kellogg's products packed during the year

1954 Sugar Frosted Flakes (Frosties) introduced

1955 Sugar Ricicles introduced; first Kellogg's TV commercial

1957 Baking-soda-operated plastic submarine given away with cereal. Cutlery promotion began with three teaspoons marked with the purchaser's initials at 3s 11d. 30,000 sets of three were ordered, but 10,000 orders a day arrived; soon half a million had been received. . . For a time Kellogg's were the country's largest cutlery retailer.

1958 Sugar Smacks and Variety introduced

1959 Special K introduced

1960 Coco Pops introduced

1961 Whole Wheat Flakes, All Stars and Handi-Pak introduced

1962 Sultana Bran introduced

1963 Bran Buds introduced; Beatles fan club badges offered on Rice Krispies packets, for 2s 3d plus two tokens

1964 Manchester factory produces over 1,000,000 lbs of cereal in a day

1973 Country Store introduced

1979 Super Noodles introduced

1980 Crunchy Nut Corn Flakes introduced

1981 Tony the Tiger celebrates his 25th anniversary

1983 Summer Orchard launched

1984 Fruit 'n Fibre and Start introduced

1985 Honey Smacks introduced

1986 Nutri-Grain introduced

Information by courtesy of Kellogg Company of Great Britain Limited

Left April 1950. Bute Town, the dockland area of Cardiff, was populated by quite a mix of cultures and was given the nickname 'Tiger Bay'. The scene is bleak but the children don't seem to notice as they play happily on the rather basic equipment, which is made from plain bare wood and metal. The youngsters on the plank-swing seem to have picked up quite a bit of speed and are hanging at an alarming angle over the unforgiving tarmac below. There is a lone adult pushing a couple of youngsters on the roundabout, but otherwise the children play unsupervised in what was described at the time as an area 'with a bad name, but a decent heart'. The playground may have been sparse, but at least it provided a focal point for the community, a central play area for all the children from the locality. Note the pushchairs, cumbersome by today's standards, and providing little protection from the elements.

Play it safe

1950 photo ©Hulton Deutsch Collection Ltd; 1993 photo and notes Julia Thorley

Right Spring 1993. Rothwell, Northamptonshire. Another 'community centre', but this time a purpose-built one. Today's children still enjoy the same equipment - swings, slides and so on - but it is more carefully designed with safety the main consideration. The bark chips on the ground are intended to provide a softer landing for the inevitable tumbles, and were the up-to-the-minute surface when laid. However, even these have now been superseded by colourful rubber-based safety surfaces that deform on impact, absorbing much of the energy of a fall; they are also slip-resistant but non-abrasive. Guidelines issued by RoSPA even discourage the use of grass directly beneath high equipment because it becomes muddy and therefore slippery in the wet, but in the dry can be as hard as concrete. Note also the safety bars around the toddler swings, and the barriers along the side of the climbing frame. Some roundabouts are even fitted with a hydraulic speed restrictor to prevent over-spinning. The modern playground is much more colourful and easier on the eye that its 1950s predecessor, but today the children are only allowed to play under the watchful eye of a grown-up.

Rails at the rectory

Fifty years ago this year a small boy caught the measles, but this time the result wasn't just spots - from it grew one of the most enduring characters in children's literature.

Christopher Awdry
tells the story of Thomas the Tank Engine and his friends

My mother probably thought it an ill wind when, in 1943, I began to develop symptoms of measles. Later experience has shown that this illness was the catalyst for at least a minor sensation in the world of children's books, but I was naturally unaware of this, being a mere three years old at the time. My father, though somewhat older, was unaware of it too, and completely unsuspecting when he made up three stories to amuse me while I lay in bed. Railways were his hobby and interest, and it was natural that he should choose a subject he knew well. We began with a rhyme:

> 'Early in the morning, down by
>   the station,
> See the little engines, all in a
>   row,
> Along comes the driver, pulls
>   the little handle,
> "Puff, puff, ssh, ssh, off we go".'

He drew a picture to illustrate the rhyme, and, presumably as an extra attraction to me, put faces on the engines' smokebox doors. It was not a new idea, for Graham Greene had used it during the '30s, but the different facial expressions drew questions:

'Why is that engine sad, Daddy?'

'Because he hasn't been out for a long time,' he replied.

'What's his name, Daddy?'

'Edward,' he said, using the first name that came into his head.

'Why hasn't he been out for a long time?'

'Because the other engines are bigger and stronger than he is, and the drivers choose them first,' he extemporised.

And so it began - and if anyone cares to reread the first page of the first book (*The Three Railway Engines*, Heinemann Young Books, £2.95 - part of the prize in this issue's competition on page 121) its relationship with the dialogue set out above will be apparent.

So Edward, Gordon and Henry were born. The stories were not written down at first, but soon came to be, in sheer self-defence on my father's part, since I quickly came to be so familiar with them that I quickly complained if they were not told as I remembered them.

There, however, matters rested for a while, and though my mother, grandmother and godmother all thought the stories worth '. . .doing something with' (though they did not know precisely what), my father, then a curate at King's Norton, a parish in South Birmingham, was unconvinced. In any case he had no idea about how one got stories published. Nothing further happened therefore until my grandmother took a hand. A distant cousin, at that time in the army but whom she thought to be connected, in civilian life, with a firm of literary agents, had asked her to put him up for the night.

'If you want me to show him the stories,' she wrote to my father, 'send them at once'.

After about twelve months of hawking them about during his leaves and being regularly turned down, this cousin met Mr Edmund Ward, a retired director of the de Montfort Press, in Leicester. Ward was disturbed at the lack of books for children, and had determined to use, during his retirement, his experience to produce a series of books for children to the best quality that he could. He had already made plans for four of his five projected books, and told our cousin that if my father would write a fourth story extracting Henry from his tunnel and providing a happy ending, then he would publish. *The Three Railway Engines* duly came out in May 1945 and proved an instant success.

In April 1946 my father moved to his first living, at Elsworth, a village about nine miles west of Cambridge. In the autumn of that year, four stories about a six-wheeled tank-engine called Thomas were published. These stories had been written at my request in about 1944, to give life to a little model engine which my father had made out of odds and ends when I recovered from my attack of measles.

But now my father was very busy in a parish much run-down after the war years and there was no time for writing. A year passed, and then, in the spring of 1948, Edmund Ward telephoned and asked for another book. The result was *James the Red Engine*, written quickly and published that September, and the following year Thomas was given a second book to himself. *Troublesome Engines* (1950) introduced a little green saddle-tank called Percy, and subsequent titles, appearing regularly each autumn, featured other engines in turn.

Children are hawk-eyed readers. By 1949 my father was receiving many letters pointing out slips here and there, and was having to concoct answers. Now you should not shatter a child's illusions by simply saying glibly that the artist has made a mistake - it might be true, but from the child's point of view is a highly unsatisfactory way of dealing with the problem. A credible method had to be devised, and gradually an imaginary island where the railway could run was developed. Mapping was sufficiently advanced in August 1951 for Edmund Ward to have a 4 ft x 3 ft relief map of the island made for my father. It still hangs on the wall of his study, to the great mystification of many visitors.

Quite early Thomas and his friends took on an even more factual form. Elsworth Rectory had a number of outhouses, one of which had already been pressed into service as a workshop. As funds became available through book sales, my father determined to achieve an ambition and build a model railway in a room adjoining the workshop. The layout was planned in the autumn of 1948, and building began the following year. Somewhat optimistically my father promised that something would be ready to show at the Church fete, due to take place in July.

'A week before the event,' my father writes, 'the only things down on the baseboard were the track and the wiring - there were no station buildings, platforms or scenic effects, and a gap in the baseboard indicated where the dock was to be. My brother, who had come to stay for a week's holiday, and I had a pretty stiff job in the limited time to spare. We worked until 11 or 12 most nights that week, but the job was done, and to our amazement and relief the line looked well and worked throughout the afternoon almost without a hitch.'

Admission money for that first showing went, of course, to the fete, but after that my father used to open the railway to visitors by appointment, and from their contributions various necessities were bought for the church: an altar frontal, Communion linen and so on. An oak wafer box, with a plaque attached indicating that Thomas the Tank Engine had helped to raise funds to buy it, was still in use until recently.

The original Thomas model, made by Rev W. Awdry for his son Christopher. *Allan Mott*

Christopher Awdry and his father signing 'Thomas' books in 1989. *Allan Mott*

The line began on a wide shelf, where Tidmouth station stood between the dock on one side and sidings on the other. Beyond was an engine shed and turntable. The main line followed the wall anti-clockwise, passing through a tunnel and curving into Knapford Junction, against the wall opposite Tidmouth. This is the junction for Thomas's branch line to Ffarquhar, which continued around the wall. The main line, however, went into another tunnel and through the wall into the workshop, where it made a wide loop back to itself. This was the 'Other Railway' and carriage sets were painted in different liveries on their opposite sides, so that they looked like different trains when they re-emerged.

The tunnel between Tidmouth and Knapford was built of old pipe lagging on a wooden frame, and must have been quite cosy inside. My father picks up the story. . .

'Later in the year,' he writes, 'I went into the railway room. . .[and] as I had not used the line for some time I sent Thomas on a test run. He passed into the tunnel, and I heard a sort of scrabbling noise. Thomas emerged from the portal, followed by a mouse, which looked round and shot back at once. Next time through Thomas had a dead-heat at the opposite tunnel-mouth - Thomas rocked and nearly came off the rails but had the best of it. The mouse retired hurt, and gave no sign when Thomas went through the tunnel a third time. He reached Knapford and reversed - I gave him full regulator and he thundered into the tunnel. This time the mouse did not wait. He shot from the opposite

Fact meets fiction: all the original Railway Series engines were based on real prototypes, but in *Stepney the 'Bluebell' Engine*, published in 1963, a real, preserved, engine appeared, former London, Brighton & South Coast railway No 55 *Stepney*, built in 1875.

'They've made a place in England called "The Bluebell Railway",' says Percy. 'Engines can escape there and be safe.' The very first train on the rescued and restored Bluebell line, near Haywards Heath in Sussex, was pulled by *Stepney* in August 1960.

In the Gunvor and Peter Edwards illustration, *Stepney* stands ahead of Duck, being regaled by the familiar figure of the Fat Controller; in the 1992 photograph, Christopher Awdry stands beside the locomotive on the Bluebell Railway. *Illustration from* Stepney the 'Bluebell' Engine *by Rev W. Awdry © William Heinemann 1963; photograph Allan Mott*

entrance, streaked along the track, dived into the dock, leaped 3 ft 6 in from the baseboard to the floor and was never seen again.'

My father (and myself too, for that matter) spent many hours in that outhouse. Gradually the stud of locomotives grew. All the engines depicted in the books are based on real prototypes, though some have had modification - Gordon has his roots in Gresley's 'A3' (*Flying Scotsman*) Class, Toby on the Great Eastern Railway's Class 'J70' 'tram' engines, and so on. Henry, as originally drawn, looked like nothing in particular (one almost wrote nothing on earth), but had been intended as a Great Central Robinson 'Atlantic', quite different in outline from Gordon. But the artist refused to believe there was any such thing, and drew a green version of Gordon. Trouble began when Henry chose to be repainted blue, as a reward for helping with Gordon's express in the final story of *The Three Railway Engines*. Readers soon complained that they could not tell Henry from Gordon. A reversion to green made no difference, and after some heart-searching in which it was even considered that Henry would have to be 'written out' of the Series, my father told the story of the only major accident the railway has had. Henry was sent to Crewe for rebuilding - he came back as a Stanier Class '5' 4-6-0 and there has been no further trouble.

The Railway Series has had six different illustrators. The first two did not last long, and C. Reginald Dalby took over for *James the Red Engine*, also redrawing *The Three Railway Engines*. John Kenney, who drew from life, took over with No 12, and the engines began to look more convincing. When he unfortunately died, after illustrating six books, the husband and wife team of Peter and Gunvor Edwards took over and saw out my father's part of the Series. Clive Spong has illustrated all my contributions, and many people have remarked how well he has managed to capture the spirit of the early books.

So the years passed, and from 1948 to 1972 a new title appeared every year. The stories became more diversified, diesel and narrow-gauge lines were introduced, until the tally had become 26 books and there were more than 30 characters. But subjects for use without repetition were, felt my father, becoming scarcer - it is his proud boast that every incident appearing in a story, however offbeat it may seem, actually happened, or could happen, to an engine somewhere at some time, and he has evidence to prove it. One was even illustrated from photographs taken by a boy travelling on the train which forms the basis for the story - 'Super Rescue' in *Enterprising Engines*. This problem, and the growing feeling that he was becoming a little stale, made him determine to stop the Series. And so, despite the publisher's blandishments, that is where everyone thought it would end.

They had, of course, reckoned without fate taking a hand. Ten years later my own involvement in the Series came about by accident, and was shortly followed by the television series which has made so many more children aware of the stories. The rest, as they say, is history. Perhaps it can truthfully be said now that the attack of measles I suffered, fifty years ago this year, was not such an ill wind after all. ●

5 sets of autographed 'Railway Series' books to be won!

As you have just read, the first 'Railway Series' book was published in 1945, and to celebrate the 50th anniversary of the telling of the original stories, we are offering as prizes five sets of the first three books, specially autographed by both Rev W. Awdry and Christopher Awdry. Each prize will consist of *The Three Railway Engines*, *Thomas the Tank Engine* and *James the Red Engine* - books to be treasured by children of all ages and, being specially signed editions, family heirlooms to be handed down through the generations!

Once upon a time. . .

How to enter

This puzzle is based on the names of children's characters from fiction and film, past and present - all you have to do is identify the names of the characters (and one author) from the clues and enter the names in the grid. When you have entered them all, their initial letters, reading across, will spell out the name of another character from children's literature, also with railway associations!

Then all you need to do is to *write this name only* - not the full answers to the ten clues - on a postcard, and send it to

Competition
Past and Present
Unit 5
Home Farm Close
Church Street
Wadenhoe
Peterborough PE8 5TE

A set of the three books will be sent to each of the senders of the first five correct entries picked at random after the closing date, which is Monday 13 December 1993. Please do not enclose any correspondence with your entry.

The names of the winners will be published in *Past and Present* No 4.

The Past and Present Prize Competition

1	2	3	4	5	6	7	8	9	10
P	A	D	D	I	N	G	T	O	N
E	L	E	O	V	U	O	R	N	E
T	F	N	N	O	T	R	E	E	S
E	I	N	A	R	K	D	V		B
R	E	I	L		I	O	O		I
		S	D		N	N	R		T

Clues

1 'Once upon a time there were four little Rabbits, and their names were - Flopsy, Mopsy, Cotton-tail, and ——.' (5)

2 Older brother of little Annie Rose in the popular stories by Shirley Hughes (5)

3 Gnasher's (in)famous young owner, sporting a red and black striped jumper (6)

4 First name of a 'fowl-tempered' character who first appeared supporting Mickey Mouse in a 1934 short (6)

5 Welsh steam engine who chuffed through a series of Oliver Postgate TV animations (4)

6 'The Tale of Squirrel ——' by Beatrix Potter (6)

7 The Big Engine, No 4 on Thomas's railway (6)

8 The name of the traction engine who features in several of the 'Railway Series' stories (6)

9 'And if anyone knows anything about anything,' said Bear to himself, 'it's —— who knows something about something. . .' (3)

10 And finally, to round off our semi-railway theme, the surname of the author of *The Railway Children* (6)

Competition rules

1 The 'How to enter' instructions form part of the rules 2 The competition is not open to employees of Past & Present Publishing Ltd or their families, or their advertising agents, printers or distributors 3 The prizes will be awarded to the appropriate number of senders of winning entries drawn after the closing date. A winning entry is one which has a correct answer to the competition in accordance with the 'How to enter' instructions 4 Entries must be received no later than the stated closing date 5 No claim is necessary; winners will be notified 6 The editors' decision is final, and no correspondence can be entered into.

Writing back

Do any the articles in this issue of Past and Present *bring back memories? Do you have any 'then and now' comparisons that you think we and our readers would be interested in? Has your job or the equipment you use changed over the years? Whatever you would like to share with us of a 'past and present' nature, we would be only too pleased to hear from you. Write to Will Adams or Julia Thorley at the editorial addresses on page 65.*

Issue 1: the verdict

Christopher F. Gledhill, who lives in Switzerland, is a staunch Past and Presenter, and was keenly awaiting the first issue of *Past and Present*, having early on taken out a subscription. By way of throwing down a whole armful of gauntlets to us the editors and you the readers and potential contributors, here are some of his comments.

I was somewhat disappointed with the outcome. What I was expecting. . .was basically the proven 'BR Past and Present' format spread over a wider range of subjects and garnished with those features which characterise a magazine rather than a book, ie readers' letters, snippets of info, regular columns, etc.

'Cumbria Railways' and 'Birmingham Trams' fully meet my expectations, of course. 'Scouting Days' comes close. Most of the others, however, are no more than potted histories with a meagre sprinkling of very dull and sterile publicity shots, devoid of all atmosphere and evocation.

It's *photos* we want. And *lots* of them. And bigger! At the moment too much valuable space is taken up with text, and too little with photos. . .

I also think you've tried too many subjects at one go. A laudable attempt, but many inevitably lack depth. . .

Finally, the cover announced 'a nostalgic look at *five* decades of change', but I didn't see very much from the 'forties!

But enough of the criticism, which is intended to be constructive and not destructive. . . I am convinced the basic concept is good. . .

Now here are a few ideas for future articles:

National Service
The milkman: daily deliveries p & p
Lyons Corner Houses to McDonalds
Odeon/Gaumont/ABC to bingo halls
Street newsvendors p & p (their pitches, too)
Cricket on the village green p & p
Rubbish collection: rag-and-bone men to bottle banks
Street lighting p & p
News Theatres at main-line stations (when and where was the last showing?)

I shall be interested to hear if I am alone in my opinions on your opening issue. Future ones will tell. I look forward to them!

Well, do you agree? We'll certainly try to include more, bigger pictures. But the content of *Past and Present* depends heavily on *you*. Tell us what you would like to see, or, better still, write the piece or take the photographs and send them in. The sooner you do, the sooner the journal will begin to reflect what *you* want to see in it!

Coventry City Centre

It was what the photos showed rather than their size than stirred Mary Elliott of Coventry to write to us following the 'Coventry City Centre' article.

What a fascinating article about my native city in your first issue of *Past and Present*. And how it stirred memories of Coventry at a time when it was struggling to get to its feet after the disastrous effects of the Second World War. I barely remember the city before the Blitz, but have vivid memories of the devastation and, as a child, lived through the raids which literally took the heart out of the city.

It made me very sad to see your excellent photograph of Broadgate. The very centre of the city was intended to be a garden island with a lovely statue of Lady Godiva in the middle - a beautiful open space of trees and flowers and lawns. All that has now disappeared in the cause of 'progress', and shops clutter the area which once brought light and air into the heart of a busy shopping town. Lady Godiva herself is now hidden beneath a hideous canopy that looks as if it has escaped from a production of 'The Desert Song', and the whole concept is a cause of much controversy in the city, which shows no signs of diminishing.

Sadly, too, much of the reconstruction of the city centre took place at a time when the fashion in architecture was for concrete boxes, and other cities, developed much later, have produced a much more pleasing aspect than ours. Now our City Council seems determined to fill every available space with retail outlets, and the lovely plans of the postwar period are fast being changed so that the original ideas have been swept away, losing so many pleasing aspects planned for us after the war.

So many nostalgic looks at the past go too far back for many of us to remember. Congratulations on choosing an era which we do remember well, and congratulations too on presenting your theme with such clarity on such nice paper to handle, and with so many interesting features. I look forward to the next issue!

More on 'the Pop'

The photograph of Allan Mott's Ford Popular struck a strong chord, both with readers and within the *Past and Present* office. John Barnett of Finchley sent us these memories.

I enjoyed most of the articles in your exciting new magazine, but the one that delighted me most was Allan Mott's piece about the Ford Popular. My own first car was that very same 1959 model. I still have the bill - the cost on the road, brand new, was £444!

The Pop was basic in all senses, but very reliable. It was also incredibly slow by today's standards. I remember vividly a drive across the Isle of Wight to catch a ferry home from our holiday. We were late and on the straight bits I exceeded 50 mph! The engine nearly burst, it seemed to me. Does anyone remember the maximum speed advertised for the dear old 'Pop'?

Past and Present editor Will Adams also has fond memories of the 'Pop'.

My family's first car was a Ford Popular, and although we, and I myself since I have been driving, have owned numerous cars since then, it is the 'Pop' that still lingers in the mind, and its unforgettable number, XHP 744. We bought a new wooden garage to house it, and I can still remember the combination of smells of fresh creosote and new car interior every time we got it out for a run - what a thrill!

For many years we went on holiday to a self-catering cottage in the Cotswolds, and the 'Pop' was loaded to capacity. My sister and I sat on and amidst blankets, pillows and picnic basket. The transistor radio slipped neatly under the front passenger's tip-up seat. When we went up hill everything fell off the dash shelf; when we went down, everything fell off the back window shelf.

Talking of hills, power to the windscreen wiper motor decreased in proportion to the work the car's engine was doing. Going uphill in a rainstorm, dad could frequently not see where he was going!

The boot lid hinged at the bottom, retained by adjustable webbing straps; if the boot was too full, these straps could be tightened to hold the lid partly open. The rear number plate could also be hinged down so that it hung vertically and could still be read. The boot key was a large, square-ended affair, not a conventional key. On our numerous picnics it was frequently lost or left at home, but a screwdriver or other tool could be called on to do the trick.

All sorts of little things about the car come to mind. A separate ignition key and pull-button to start the engine. A handbrake with an L-shaped end that came horizontally out from beneath the dash shelf. Plastic seats that froze bare legs in winter and burned them in summer.

One useful aspect of the 'Pop' was that the bonnet was in two halves, hinged in the middle and lifting up from each side. Running beneath and parallel to the centre

Will Adams and his sister Judith pose in front of the family's pride and joy, XHP 744, circa 1960.

of the bonnet was a metal rod. When we'd been on a picnic near water, and inevitably got our sandals wet, their straps could be buckled over this rod and the shoes left to hang during the journey. When we arrived home they would be dry!

The 'Pop' was our pride and joy for over ten years, until it finally expired in 1968. We then upgraded to an unbelievably luxurious Singer Gazelle (1401 RW) - *four* doors, front bench seat, wooden facias, and - oh luxury - a heater! Rolls-Royce nothing!

Gentleman of the press

The feature in Issue 1 describing the changes in the printing industry brought back memories for Edward Bodell of Cheshunt.

Past and Present is a treasure trove of memories. I particularly enjoyed the article 'Hot metal to microchips' and found it to be technically correct, especially the statement about modern-day typography falling into untrained

hands with the consequent falling of standards. It is with this notion in mind that I have created my own modern composing/reading room on a desktop, with the help of an Apple-Mac computer, colour scanner and 600 dpi high-quality laser printer. Currently I am establishing a desktop publishing service offering the traditional typographical approach, while observing all the accepted practices from the 'hot metal' days, with the up-to-date methods of graphic and design origination.

I found the picture of the compositor correcting imposed pages most interesting. I am surprised that wooden quoins were used to secure imposed pages in a chase (most risky!). I used metal expanding quoins which opened outwards by means of a key being inserted and turning a cam to open two opposing wedges. Wooden quoins in my day were only used to secure type on a galley when a wooden side-stick was placed alongside the type and the wooden quoins were knocked into position with the mallet and 'shooting stick' to prevent the lines or type from sliding. When by accident type ended up in a jumbled heap, the news would travel around the composing room because of the torrent of expletives. Everyone would gather round to view the compositor's misfortune and call out 'printer's pie!'.

Scouting Days: Allan Mott's pictures in Issue 1 reminded *Past and Present* Publisher Peter Townsend of his own scouting days with the 18th Swindon troop. 'I well remember the annual camp being the high spot of the scouting year. We were quite adventurous as a troop, visiting Guernsey and the International Scout Chalet at Kandersteg in Switzerland during my time. I wonder if the troop is still active and visiting such faraway places - or perhaps even further afield? I have been rummaging through the family photos and came across this group shot taken just before departure for Switzerland - where are they now, I wonder? Unlike Allan's troop we were, I recall, proud to wear the traditional Scout hat, a point we lost no time in teasing our fellow Swindon troops about - to us they were the real thing! A steaming kettle to press the rim of the hat was an essential, steam irons being rare at that time. *Wiltshire Newspapers Ltd*

'Our age and generation'

Anthony Phillips of London brought back this timely pair of photos from a recent trip to Boston, USA.

Spring 1946 in the North End neighbourhood of Boston, Massachusetts, and the buds on the English elm trees are just beginning to appear. A rather handsome young man seems to stand out head and shoulders over a small gathering of local residents who are predominantly of Italian immigrant stock. He talks earnestly of the benefits and responsibilities of living in a free society, of civil rights, of religious perspectives, and of labour policies which would provide jobs for all.

This is John Fitzgerald Kennedy at 29, an Irish-American Catholic campaigning for the United States Congress, one of just 435 local representatives to take their seats in the nation's capital, Washington DC. His listeners are youngish and enthusiastic. A shoe-shine boy (*extreme lower left*) ponders how on earth this stylish gentleman from the leafy Boston suburb of Brookline, a real Harvard University scholar, an ex-Naval office,

could possibly help. They were both staunch Democrats though.

Soft, wide-brimmed felt hats are in vogue for the men, and the women wear warm, sensible knee-length topcoats and court shoes. The bicycle in the foreground? A Schwinn. Its young owner has saved hard to get a real sheepskin saddle to give a softer ride than the original leather.

John F. Kennedy stands beneath the Paul Revere statue, honouring an American patriot who in 1775 made a legendary night ride to Lexington and Concord to warn of the advance of the British redcoats during the revolutionary war. In the background is the old North Church built in Christopher Wren style with 'wedding cake' spires. In 1723 this was the tallest building in Boston.

JFK, son of prosperous Joseph P. Kennedy and Rose Fitzgerald of East Boston, is destined to become the nation's 35th President. Milestones during his term of office will be the Bay of Pigs, the Berlin Wall, the Cuban missile crisis, the Alliance for Progress, the Peace Corps, economic policies, and explorations into space. He will marry Jacqueline Lee Bouvier, a French-American socialite, on 12 September 1953, and there will be two children, John F. Kennedy Jr and Caroline B. Kennedy.

The second photo was taken in the

autumn of 1992, and the rich foliage, though turning to vivid red, gold and yellow, obscures the old North Church. One of the many visitors to Boston passes within a couple of feet of where John Kennedy stood 46 years earlier. His head is turned to the right as though in acknowledgement.

The granite setts in the square, where the local residents play chess in the summer, have been replaced by warm red brick laid flat, giving a romantic effect. The lamp column to the right of the statue looks similar but is darker and more slender; the globe looks the same. The trees are stronger and sturdier, and the heavy wrought iron tree-guards are no longer necessary. The stone seats around the square are the same. The Paul Revere statue, the centrepiece, is one point of interest on the 'Freedom Trail', a mile-and-a-half walk through Boston's historic streets. It was here in Boston that the Declaration of Independence was signed in 1776.

And JFK? On 22 November 1963, just 30 years ago, his life was taken by a bullet fired by assassin Lee Harvey Oswald, whilst campaigning in Dallas, Texas, for further presidential office.

1946 photo courtesy of The John Fitzgerald Kennedy Library, Boston, USA

Now then . . . ◇ ◇ ◇

The Past and Present guide to museums, collections and societies where the past can be remembered and enjoyed today.

Tram-ride to yesterday

Those of you who share Alan Bennett's fond memories of tram rides, whether in Leeds or in any other British city, can do no better than pay a visit to the National Tramway Museum at Crich, near Matlock. The museum is home to over 40 steam, horse-drawn and electric trams, many of which have been painstakingly restored by staff and volunteers. The trams on display are all original, having been saved from the scrap-heap by far-sighted enthusiasts conscious of the need to preserve and run a much-loved part of our transport heritage.

There are many exhibitions to look at, and authentic buildings and street artefacts surround the central cobbled period street. There is also a delightful woodland walk, and displays about lead mining in the area.

Visitors are encouraged to take as many tram journeys as they wish, and there are a number of different trams in service running every few minutes. The two-mile round trip starts off along the cobbled main street, heads off into shady woodland and emerges on a hillside giving breathtaking views along the Derwent valley.

And if you arrive in a classic car or bus, the driver and one passenger over 25 years old will be admitted free provided that the vehicle is parked in the museum tramway street for at least two hours!

* One of the trams at Crich is Leeds tram No 180, so you can experience 'Leeds trams past and present' simultaneously!

> **National Tramway Museum, Crich, near Matlock, Derbyshire (0773 852565**
> **Adults £4.20, Children £2.40.**
> **Open daily 10 am-5.30 pm. Last admission 4 pm.**

Loading at the Town End terminus at the National Tramway Museum, Crich, is car No 106, one of the first trams to operate on the London County Council tramways in 1903. It was extensively restored by Tramway Museum Society members in London before being moved to the Museum in 1983.

The building facade in the background is from Derby Assembly Rooms, and was all that survived a disastrous fire in 1963. It was dismantled and moved, stone by stone, to be re-erected at the Museum site in 1972. *National Tramway Museum*

National Waterways Museum, Gloucester

Fifty years ago, Gloucester Docks, though already in decline from their former importance as one of Britain's largest inland ports, with a convenient outlet to the area at Sharpness, was nevertheless still commercially viable.

As I recall, when my wife and I spent a couple of happy days there just after the end of the Second World War, the docks and the canal which served them handled barges carrying timber and miscellaneous cargo, as well as small oil tankers and diesel-power freighters. Still clear in my memory is the occasion when two friendly tugmen invited me to pass the forbidding 'No Trespassing' notice and pose for a snapshot on their vessel!

Today the area on which our visit was concentrated has taken on a new life with the establishment of the National Waterways Museum in one of the most magnificent of the Victorian warehouses that still delight the eye of the connoisseur of things past. This is the seven-storey Llanthony Warehouse, recognisable today as the building we admired on the basin known as the 'Barge Arm', and which in 1946 was still fulfilling its commercial role. Its main purpose now is to provide moorings for the museum's floating exhibits, including such historic craft as their huge No 4 Steam Dredger, canal narrowboats *Northwich* and *Oak*, and a rare concrete barge, still undergoing restoration after being 'rescued' from the Severn.

Ashore, visitors can see the steam crane, built by Balmforth of Leeds around 1880, and now restored to full working order. These and other exhibits are the successors to the craft and shore-based equipment that we saw in action 50 years ago, and which today stir memories of those days.

Not available to us then but on offer to today's visitors to the museum are cruises of varying length on board *Queen Boadicea II*, on the canal and the Severn depending on the programme of events. Built in 1936 and measuring 65 feet long, *Queen Boadicea II* herself has undergone very little change during her

more than 50 years' service, apart from refurbishments of the fittings, the installation of new powerplant and other examples of upgrading. Employed as a pleasure launch in various estuaries along the south coast, the vessel was commandeered during the Second World War, her most notable achievement being ferrying troops from the Dunkirk Beaches after the fall of Belgium. She survived enemy aircraft action and reached port safely so loaded with survivors that, so it is said, some were accommodated in the boiler room!

The present Llanthony bridge which spans the Gloucester-Sharpness canal is the latest of three which have spanned the dock exit, and was erected in 1972. When we first visited the docks in 1935, a single-leaf bridge carrying a railway line and built in 1862 was itself a replacement of a previous two-leaf wooden swing-bridge. Today the bridge carries a continuous stream of private cars and commercial vehicles, making it almost impossible to stand on it and take photographs of the Barge Arm as we were able to do without fear of interruption in 1946.

How appropriate it is that perhaps the most important change which has taken place at Gloucester docks over the past 50 years should have been the establishment of the National Waterways Museum in the still surviving Llanthony Warehouse. Entering through a re-creation of a lock chamber, visitors can re-live the story of Britain's waterways through workable models and machinery, archive films, sound recordings, contemporary photographs and drawings, countless artefacts, a huge working diesel engine, together with a replica maintenance yard where traditional crafts such as blacksmithing, woodwork, rope fender making and boat painting are demonstrated.

Eric Ford

The National Waterways Museum housed in the old Llanthony Warehouse at Gloucester Docks. The narrowboat *Northwich* is moored alongside. *National Waterways Museum*

The National Waterways Museum, Llanthony Warehouse, Gloucester Docks, Gloucester GL1 2EH (0452 307009).
Adults £3.95, Children and senior citizens £2.95. Family ticket £9.50.
Open daily from 10 am to 6 pm (5 pm in winter). Last admission one hour before closing.

Relics of the past

Collectors' Corner, British Railways' home for retired railway relics, is a wonderland for the railway enthusiast, collector or modeller, child or adult. If BR no longer needs it, collectors can probably find it here, be it signalling equipment (30p-£120) or publications (25p-£12), timetables (£1.25-£2) or maps (20p-£1.50), lamps (£2-£48) or clocks (£5-£500), or any of hundreds of other items. If it's no longer in service, it's probably in Collectors' Corner.

Collectors' Corner opened at Cobourg Street, near Euston Station, 22 years ago. Bob Ballard, shop director, says, 'Today you find a lot more collectors than when we first began doing business. We helped to start the trend, and now we're perceived by collectors as the main supply point.'

Ballard explained that in the 1960s the Beeching Report 'cut the size of the railway. Because of all these closures, much material became available. And we were also switching from steam to electric so locomotives' name and number-plates were becoming available.'

Rather than just scrap the material, BR started auctioning it off. 'They had auctions all over the country. The General Manager of the London Midland Region decided that there was a lot of money to be made, so he asked his stores controller to set up these sales as an ongoing project.' They found that items were coming into the offices on a regular basis. So they started going to depots and private preservation sites to assess the material and found it was very remunerative.

Ballard continues: 'In about 1970 a BR employee who had lost his job with the railway used his redundancy payment to buy a shop in North London. He set up a business buying and selling these same railway relics.'

A controller's staff member, seeing this, suggested that if he could do it and make a business out of it, perhaps BR could do the same. 'It didn't cost the railway anything because the stuff was theirs to begin with. So Collector's Corner started in December 1969 and we've been here since.'

Customers at Collectors' Corner are not just railway buffs, but collectors in general. 'The buying public is pretty much the same today as it was in the beginning - a mixture of enthusiasts, regular collectors, interested members of the general public. A lot of architects come to us for decorations for pubs, hotels, etc.'

As we spoke, Ballard was interrupted by the phone. Afterwards he told me that the caller had been the director of the Royal Shakespeare Company at Stratford-upon-Avon. He wanted a two-wheel wooden luggage trolley. For theatricals Collectors' Corner generally

hires material out at a charge 10 per cent of the value of the item per week.

T. Bruce Tober

Railway relics at Collectors' Corner come in all shapes and sizes, from nameplates and ironware to tickets and labels. Here's a cabinet full of highly collectable railway silverware. *T. Bruce Tober*

Collectors' Corner, Cobourg Street, London NW1.
Open six days a week, Monday to Saturday, 9.00-4.30. Catalogue mail-order service also available, supplemented bimonthly and obtainable by sending £3 to Collectors' Corner, Euston Station, London NW1 2HS. For further information call 071 922 6436.

LISTINGS

As well as the nostalgic attractions described in detail here, there are many more places around Britain where the past is alive and well! Here is just a selection of the wide range of subject areas on offer - details are, of course, subject to change, so please check before planning your visit. We hope to feature some of these attractions and others in more detail in future issues of *Past and Present*.

Photocopies of lists that have appeared in previous issues of the journal are available by sending a large stamped, self-addressed envelope to one of the editorial addresses on page 65.

If your museum, collection or society would like to be featured in *Past and Present*, why not drop us a line to the same address?

Avon

Museum of Costume, Assembly Rooms, Bennett Street, Bath
Tel: 0225 461111
Extensive collection devoted to fashionable dress for men, women and children from the late sixteenth century to the present day.
Open: Daily March-October 9.30 am-6 pm (Sunday 10 am-6 pm). November-February 10 am-5 pm (Sunday 11 am-5 pm).
Admission: Adults £2.40, children £1.50.

Buckinghamshire

Milton Keynes Museum of Industry and Rural Life, Stacey Hill Farm, Southern Way, Wolverton, Milton Keynes
Tel: 0908 319148
Industrial, agricultural and domestic items including tractors, farm implements, stationary engines, lawnmowers, photographic equipment, printing and telephones.
Open: 28 April-31 October, Wednesday-Sunday 1.30-4.30 pm. Bank Holiday Monday 1.30-4.30 pm. Also Easter Sunday and Monday 1.30-4.30 pm.
Admission: Adults £1.25, concessions 75p.

Essex

Mark Hall Cycle Museum and Gardens, Muskham Road, off First Avenue, Harlow
Tel: 0279 439680
Bicycles from 1818 to the present, plus a large of collection of lamps, tools, saddles, pumps and other accessories.
Open: Sunday-Thursday 10 am-5 pm.
Admission: Adults £1, children and senior citizens 50p.

Gloucestershire

The Robert Opie Collection at the Museum of Advertising & Packaging, The Albert Warehouse, Gloucester Docks, Gloucester, GL1 2EH
Tel: 0452 302309
Some 300,000 items of advertising and packaging - packs, tins and bottles that have filled the family shopping basket over the years.
Open: Tuesday to Sunday 10 am to 6 pm; open every day during the summer; closed at 5 pm during weekdays in winter. Open all Bank Holidays except Christmas Day and Boxing Day.
Admission: Adults £2.50, students and senior citizens £1.95, children 95p, family ticket £5.95.

Hampshire

National Motor Museum, Beaulieu
Tel: 0590 612123 or 0590 612345
Over 250 vehicles and displays which vividly illustrate the history of motoring. Admission to motor museum also includes Palace House and grounds, the abbey and exhibition, and vouchers for rides and drives.
Open: Daily 10 am-6 pm.
Admission: Adults £7, children £5, concessions £5.50, family ticket £22.

Lothian

Museum of Flight, East Fortune Airfield, near North Berwick, Edinburgh
Tel: 062088 308 or 031 225 7534

Over 30 aircraft and aero engines, including a Comet 4C and Vulcan.
Open: April-September 10.30 am-4.30 pm daily.
Admission: Free.

Staffordshire

Bass Museum, Visitor Centre and Shire Horse Stables, Horninglow Street, Burton on Trent
Tel: 0283 511000
7,000 years of ales and beers, covering industrial and social history of brewing. Special displays include a working railway model of Burton, glass and ceramics associated with beer drinking, and Edwardian bar and vintage cars.
Open: Daily 10.30 am-5 pm (last admission 4 pm).
Admission: Adults £3.45 (which includes a half pint), children £1.85, family ticket £9.25.

Suffolk

Museum of East Anglian Life, Stowmarket
Tel: 0449 612229
Large open-air museum in a riverside setting, with displays on agriculture, crafts, social life and industry.
Open: April-October, weekdays 10 am-5 pm, Sunday 10 am-5 pm.
Admission: Adults £3.50, children £1.60, concessions £2.50.

West Midlands

Museum of Science and Industry, Newhall Street, Birmingham
Tel: 021 235 1661
Items of general industrial and scientific nature including steam engines, machine tools, small arms, aircraft and scientific instruments.
Open: Monday-Saturday 11 am-5 pm, Sunday 11 am-5.30 pm. Closed Christmas Day, Boxing Day and New Year's Day.
Admission: Free.

Final thoughts . . .

John Lucas *News at the grassroots*

Journalist John Lucas worked for more than 30 years on The Daily Telegraph, *as sub-editor on the* Sunday Telegraph, *chief sub-editor of features for* The Daily Telegraph *and ultimately as chief sub-editor of the Telegraph's 'Weekend' section. He also wrote - and continues to write - many feature articles. Here he recalls how his career started in the 1950s.*

Local papers have not always thudded weightily through the letterbox as they do these days. In the 1950s, for example, they had few pages - only six or eight - and correspondingly few openings for aspiring journalists.

So it took me two years to get taken on. And having spent three years in the army, and at 23 a latecomer to journalism, I had much to learn. But at least I knew shorthand and typing. South Hertfordshire was in effect a pleasant extension of suburbia and I had an urban district of about seven square miles to cover on my motorbike. You worked at the grassroots of journalism here. Sometimes I think national journalists have it easy; if they had to face their readers, as local newspapermen do, there'd be less sheltering behind words. On my paper, I even had to go round and apologise to a bride's mother because in a wedding report I'd forgotten to mention her daughter's orange-blossom headdress!

The 'local' was a newspaper of record then, not a tabloid entertainment sheet desperate for higher sales as it so often is now. In the line of duty you covered every event, from district council meetings to inquests, from presentations and unveilings to obituaries, school speech-days and sports, meetings of old people's welfare committees - and church plays.

There were hilarious moments. . . For example, the play in which someone really did bring the house down. On stage, they turned a door handle, which jammed, shook it, and the entire plaster-board wall of scenery on one side of the stage collapsed. They should really have kept that scene in.

There were the poignant moments. . . I called on one couple for their golden wedding reminiscences. In their late seventies they were still clearly devoted. But the next time I saw the wife she was a forlorn widow giving evidence at her husband's inquest. He had gassed himself. 'He was so worried about the state of the world,' she said tearfully. I was almost in tears myself at seeing her there, bewildered, disconsolate, only months after their day of happiness.

You would call routinely on the police and fire brigade in search of news tip-offs, and the clergy would sometimes yield up a paragraph, and sometimes a good story. One cleric told me of the parishioner who called at his house one Monday morning and said, 'Sorry, vicar - I meant to put a penny in the plate yesterday, but I accidentally put in half a crown. Can I swap, please?' That paragraph appeared in the *Daily Express* and *Daily Sketch*.

Local newspaper editors were often also part-time correspondents for the national papers. If one of his staff had a good story, he telephoned it over, quoting the editor's name, and the payments would all be shared out once a month. It was an added incentive to scent out good stories.

There were several local murders on our patch. Once, a husband killed his wife with an axe, and I recall the senior reporter, a Lancashireman, telephoning the story to every London paper and the two home news agencies - about a dozen all told. 'A bloodstained hatchet was produced in court today. . .' I can still hear that dreadful litany as I typed out flower show results. It seemed to go on for hours.

Calling on the bereaved for obituary reports was a regular task. One day a widow took me into the back room to see her husband in his coffin. I duly paid my respects, then at the door when I left, she pressed sixpence into my hand. A tip! Heavens, how many reporters ever got that!

As a local reporter, you were seen at most local events, and you stood or fell by your reports. One week, though, seeking the local angle on a national census, I proved that many local shops were losing a lot of business to a neighbouring town, which people preferred, and my street interviews confirmed as much.

A few days after my report appeared, I found myself sitting at the local Chamber of Commerce meeting, taking a verbatim note of the chairman's attack on myself as author of the offending news item. It duly appeared on the front page.

You worked long hours then: Monday to Friday and Saturday morning, with Friday afternoon off. On Saturday afternoon you covered perhaps three garden fetes, with a concert in the evening. Often you would write it all up on Sunday, ready to gather more news on Monday. And there would be perhaps two evenings a week spent reporting committee meetings.

But it wasn't all a chore. I once fell in love with a beautiful local florist, and enjoyed many romantic Monday afternoons at the back of the shop among the wreaths and bouquets. How far away was the clatter of typewriters and the tang of printer's ink!